T0321581

UNIVERSAL MEASUREMENTS

HOW TO FREE THREE BIRDS IN ONE MOVE

UNIVERSAL MEASUREMENTS

HOW TO FREE THREE BIRDS IN ONE MOVE

DIEDERIK AERTS
Brussels Free University, Belgium

MASSIMILIANO SASSOLI DE BIANCHI
Laboratorio di Autoricerca di Base, Switzerland

World Scientific

NEW JERSEY • LONDON • SINGAPORE • BEIJING • SHANGHAI • HONG KONG • TAIPEI • CHENNAI • TOKYO

Published by

World Scientific Publishing Co. Pte. Ltd.
5 Toh Tuck Link, Singapore 596224
USA office: 27 Warren Street, Suite 401-402, Hackensack, NJ 07601
UK office: 57 Shelton Street, Covent Garden, London WC2H 9HE

Library of Congress Cataloging-in-Publication Data
Names: Aerts, Diederik, 1953– author. | Sassoli de Bianchi, Massimiliano, author.
Title: Universal measurements : how to free three birds in one move /
Diederik Aerts (Brussels Free University, Belgium), Massimiliano Sassoli
de Bianchi (Laboratorio di Autoricerca di Base, Switzerland).
Description: Singapore ; Hackensack, NJ : World Scientific Publishing Co. Pte. Ltd., [2017] |
Includes bibliographical references and index.
Identifiers: LCCN 2017010482| ISBN 9789813220157 (hard cover ; alk. paper) |
ISBN 9813220155 (hard cover ; alk. paper) | ISBN 9789813220164 (pbk ; alk. paper) |
ISBN 9813220163 (pbk ; alk. paper)
Subjects: LCSH: Measure theory. | Measurement. | Probabilities.
Classification: LCC QC174.17.M4 A37 2017 | DDC 530.801--dc23
LC record available at https://lccn.loc.gov/2017010482

British Library Cataloguing-in-Publication Data
A catalogue record for this book is available from the British Library.

Cover illustration: Luca Sassoli de Bianchi

Desk Editor: Ng Kah Fee

Typeset by Stallion Press
Email: enquiries@stallionpress.com

Printed in Singapore

Dedicated to the memory of Bart D'Hooghe, who from "behind the scenes" guided us towards the possibility of this collaboration, whose fruits are also his.

Preface

What do these have in common: an age-old paradox of probability theory, the central problem of quantum mechanics and the key issue of the modeling of human cognitive processes? Simply put, all three of these problems can be addressed, and in part solved, by exploiting the same fundamental notion: *universal average* or *universal measurement.*

The purpose of this publication is to take the reader on a fascinating adventure, an adventure which the authors had the pleasure to share during their recent collaboration and which allowed them to find the solutions that will be presented, explained and illustrated here.

The text is set to appeal to all fans of science and requires no specialist technical knowledge but only, of course, a bit of general scientific culture and a lot of curiosity. On the other hand, since we will present innovative notions that are still not well-known among professional researchers, its reading will also be beneficial to experts, because they can complete their understanding by consulting the technical articles listed in the bibliography. Lastly, the text is addressed to teachers, who we hope will find here many hints to enrich their courses and stimulate the young minds of future researchers.

Before we begin, we think it is useful to explain the origin of the collaboration that has led to the results we present in this long volume (you can certainly skip this preface, or read it at a later time, because its reading is not required for a proper understanding of the main text). We have thought it wise to use the first-person perspective, so as to avoid the inconvenience and artificiality of a text where the authors have to refer to themselves in the third person. There is no need to worry if the meaning of some of the concepts evoked in this preface is not quite clear, because they will be explained in detail in the subsequent sections.

MASSIMILIANO: In July 2013, I was working on a specific theoretical notion, *robustness*, as a possible key to understand the nature of the mysterious quantum measurements. My hope was to be able to characterize the quantum measurements as measurements that, among all possible measurements, were maximally robust. The idea was that these were actually selected by the experimenters, when adjusting their instruments, considering that the more robust the measurements were, the better guarantee they would offer in terms of the *reproducibility* of the experimental results.

To explore this idea, however, I needed a sufficiently general theoretical framework to describe the different types of possible measurements, in addition to the purely quantum ones; this was necessary to be able to calculate, and then compare, the degrees of robustness of the different measurements and thus succeed in putting in evidence that the quantum ones were indeed the most robust.

I knew that a theoretical framework of this kind had been proposed in the 1980s by *Diederik Aerts*, a Belgian physicist whose work I knew, because he was one of *Constantin Piron*'s students (see Figure 0.1). Piron co-founded the so-called *Geneva school of quantum mechanics* (afterwards better known as the Geneva–Brussels school of quantum mechanics, precisely because of the important contributions made by Aerts and his collaborators).

I knew Piron well, because in 1990 I was his assistant for about a year. His research was all focused on the attempt to reconstruct the foundations of physical theories, starting from a very general axiomatic and operational approach. More precisely, starting from theoretical notions as simple as those of *experimental test*, *property* and *state*, he was hoping to succeed in deriving both quantum physics

Fig. 0.1 C. Piron.

and classical physics as special cases of a wider and more general theory. Piron obtained some important results, but Aerts was able to bring this program to completion by identifying the physical content of some of the axioms necessary to characterize a quantum theory, and by understanding that the presence of these axioms introduced an irreducible structural limitation into the theory itself, which therefore could no longer be considered a *complete* theory (for example, because it proved incapable of describing systems that are experimentally *separated*), as most physicists instead believed, and still believe today.

On the other hand, a more general theory, not subjected to the structural limitations associated with these specific axioms, could be considered complete (or more complete) in the sense of "being able to describe all processes and conditions that are in principle observable in nature, both in macroscopic and microscopic domains."

I remembered the admiration with which Piron told me about the progress made by Aerts, and having had the occasion, in more recent times, to read and comment on some of his works, it was natural for me to start from his approach in my attempt to formalize my argument of robustness of the quantum measurements. More exactly, I became interested in the particular approach called the *hidden-measurements approach*, developed by Aerts to model the different possible behaviors of physical systems, whether classical, quantum or *quantum-like* (intermediate).

As we will explain in the text, the term "hidden measurements" refers to the fact that not always, in the course of an experiment, the interaction between the measuring instrument and the measured entity can be considered predetermined, and in that sense it would remain "hidden," because "not known," and depending on the manner in which the interaction is selected each time, different kinds of measurements can be obtained.

In the 1990s, in the context of his axiomatic studies, Diederik Aerts had already asked the question about the specific nature of quantum measurements, and more precisely about the nature of the probabilities they are associated with (called *quantum probabilities*, to distinguish them from the *classical probabilities*, which emerge from the *Boolean logic*). Indeed, because he had observed in his hidden-measurements approach that different typologies of measurements were *a priori* possible, and that the classical and quantum ones were only particular cases, it was only natural for him to wonder what could be the mechanism that, in the measurement processes of microscopic entities, made that it was precisely the quantum measurements, with their specific probabilities, which emerged, rather than other typologies of measurements.

To this question Diederik gave a fascinating answer, although decidedly speculative: according to his hypothesis, the quantum measurements were authentic *universal measurements*, in the sense that they were the result of a gigantic average over all possible types of measurements. Therefore, they expressed a level of *indeterminism* much more profound and fundamental than what one might initially suspect — provided, of course, that his hypothesis proved correct.

Now, when I formulated my hypothesis, within the framework of the hidden-measurements approach, that quantum measurements were such because they were maximally robust, my first impression was that there was a conflict between this explanation and the one proposed by Diederik, according to which the quantum measurements were instead the result of a universal average.

This led me to write a first draft of a short article on the issue, and in it I also expressed my criticism of the validity of the idea of universal measurements. Before making public the final version of the article, by uploading it to the main repository of scientific articles (arXiv.org), I sent it to Diederik, to stimulate him to respond to my criticism of his hypothesis.

He answered promptly, expressing a great interest in the possibility of identifying other notions, such as that of robustness, that would help explain the emergence of quantum measurements and quantum probabilities, but inviting me at the same time to observe a measure of caution. As he saw it, in fact, it was not necessary to oppose the notion of robustness to that of universal measurement: both were worthy to be deepened, and had their own interest, but were not for this to be considered as competing. It could very well be, he said, that quantum measurements are universal *and* at the same time maximally robust.

Diederik's response triggered a long and fruitful exchange of messages, with myself trying to better understand the subtle notion of universal measurement, but remaining rather skeptical about the suggestion to not only assign physical meaning to the notion, but also find a quantitative correspondence with the quantum measurements. This was mainly because, to do so, one had to solve a very thorny issue, unsolved since 1889: the famous *Bertrand's Paradox.*

Indeed, it was this paradox that made it impossible for the notion of universal measurement to be formulated in a sufficiently clear and comprehensive manner, at the time when Diederik was trying to define it in the 1990s: any attempt to do so with greater mathematical precision was immediately frustrated by the difficulties involved in this famous problem, which lay at the foundations of *probability theory.*

At the end of this initial and lengthy exchange of ideas, Diederik invited me to continue our discussion, but this time with the aim of summarizing our conclusions in a joint article. I accepted his proposal with enthusiasm, also because, the more I discussed this matter with him, the more I became aware of the fact that I had not really understood the nature of universal measurement.

Our subsequent discussions evolved more or less as follows. Every time I tried to highlight what I thought could be inconsistencies in his notion of universal measurement, Diederik would reply by turning upside-down all my arguments, and every time I would find myself a bit more confused. While I continued to believe there was a problem in the notion, my conviction waned.

I should add that in previous years Diederik had already attempted to resolve the problem of this particular universal average from different angles, partly through his temporary collaboration with a mathematician who was an expert in probability theories, or by inviting one of his doctoral students to collaborate. However, with every attempt, intractable issues emerged related to the very foundations of mathematics.

For example, starting from set theory to define a universal average required choosing a method for defining the set of measurements on which to perform the average. In this way, however, the result of the average, which itself is a measurement too, depended on the method chosen, so that one had to again choose a method for averaging over the different ways of choosing a set, and so on, thus giving rise to a disastrous infinite regress.

When he explained to me all these difficulties, not to say impossibilities, I remember that, seeing all my perplexities, Diederik said something that left a lasting impression on me: *that we should not make physics subordinate to mathematics*! In our case, the notion of universal average could very well possess a deep physical meaning, although it may not be possible to attach to it a precise mathematical meaning.

I cannot remember if it was because of this statement of his that impelled to try something different, but anyway, rather than addressing the problem directly, which seemed to me impossible to do, I attempted to build an ultra-simplified "toy model" that would greatly facilitate dealing with the problem. Actually, my initial motivation to do so was that of finding a counter-example. Indeed, I thought, if the idea behind the universal measurements would prove to be false in a simplified context, *a fortiori* it had to be false in the general context.

But to my surprise, the toy model validated Diederik's thesis, providing the starting point that allowed us to generalize the approach and obtain a complete solution to the problem of the equivalence between universal and quantum measurements. In short, Diederik had been right from the beginning regarding the validity of the notion and its full compatibility with the notion of robustness: quantum measurements are *both* universal measurements (or rather, a particular type of universal measurements) *and* at the same time maximally robust measurements.

Even so, my initial disagreement had been helpful in that it created a constructive dynamics that subsequently led us to a complete solution to the quantum measurement problem. But not only that, because thanks to the connection between universal measurements and Bertrand's paradox, we quickly realized that the same notion (of universal average) also enabled us to solve this important problem, which had undermined the very foundations of probability theory for over a hundred years.

Lastly, Diederik being one of the pioneers of the so-called *quantum cognition*, a multidisciplinary field of study in which the quantum mathematics is used to model the human cognitive processes, we also realized that our result offered a very convincing explanation of why the quantum probabilities were so effective in modeling human cognitive processes. It was simply because cognitive scientists, in their experiments, carried out, without realizing it, universal measurements, and that quantum measurements are universal measurements. In short, "*in one universal move we had freed three big birds*!"

DIEDERIK: It was a fortunate day when Massimiliano contacted me, presenting me with his skepticism about the notion of *universal measurement*, and about the conjecture discussed in some of my articles, i.e., that *quantum probabilities* would be universal averages over all possible types of non-classical probabilities, and could hence be interpreted as the probabilities of a sort of *first-order non-classical theory*.

When I formulated this conjecture, the analogy I had in mind was how "the first term in a Taylor series is a *first-order approximation* to any holomorphic (i.e., infinitely differentiable) function", and that "this role would be played in the realm of *non-classical probability models* precisely by the quantum probabilities." If such a conjecture was shown to be true, it would explain the success of quantum probabilities in a similar way as linear functions — which are this first term in the Taylor series of holomorphic functions — are successful in modeling situations where the functional relation is unknown, as is done in "linear regression approaches in statistics."

At the time when Massimiliano contacted me and we started to discuss about this conjecture, I had not reflected about it for more than a decade, so my mind was engaged in very different issues. However, right away it was triggered again by Massimiliano's remarks. As he points out in his preface, I had tried very hard — initially by collaborating with a mathematical probabilist and later proposing the problem as a PhD subject to one of my students, *Bart D'Hooghe* (see Figure 0.2) — to find proof for the conjecture, each time stumbling on aspects of Bertrand's paradox.

What personally made me keep believing in the truth of the conjecture was my *physical intuition* that "if such a bunch — not to call it a set, because then Bertrand's paradox strikes — of non-classical

Fig. 0.2 B. D'Hooghe.

probability models exist in physical reality, then some sort of average needs to show up statistically in this physical reality, as a consequence of our lack of knowledge about which of the non-classical probability models governs in every specific situation." From this belief, most probably, resulted also the somewhat challenging statement that *we should not make physics subordinate to mathematics*, which triggered Massimiliano to look for a toy model.

Anyhow, "the rest is history," as they say. Indeed, Massimiliano, much to my admiration and amazement, after his toy model confirmed the conjecture, found the steps towards a general proof of the conjecture, and this made it possible to indeed "*universally free these three birds.*"

Diederik Aerts and Massimiliano Sassoli de Bianchi

Contents

Contents

Chapter 1

Bertrand's Paradox

As indicated in the title, and in the preface, the central theme of this booklet is the notion of *universal measurement* (or *universal average*). Our goal is to explain how this fundamental notion can solve three major problems, in three fields of knowledge: the measurement problem (or the problem of observation) in quantum theory, the Bertrand's paradox in probability theory, and the problem of the unreasonable effectiveness of quantum probabilities in experimental psychology. We will start by addressing the famous Bertrand's paradox, which since 1889 has been a threat to the validity of the so-called *principle of indifference*, on which the entire probability calculus is based. *Joseph Louis François Bertrand* (1822–1900) was a famous French mathematician (see Figure 1.1), who contributed to many fields of human knowledge: number theory, differential geometry, probability theory, economics and thermodynamics. In his celebrated book on probability theory (Bertrand, 1889), he enunciated the following problem:

If we draw at random a chord onto a circle, what is the probability that it is longer than the side of the inscribed equilateral triangle?

Bertrand gave three different answers to his seemingly innocent question, associated with three distinct values for the probabilities in question. And because these three answers each appears to be based on a perfectly correct and logical reasoning, Bertrand's problem was subsequently qualified as a paradox (by Henri Poincaré).

Fig. 1.1 J. L. F. Bertrand.

The *first solution* proposed by Bertrand consists in choosing an arbitrary point on the circle, considering it as one of the vertexes of an inscribed equilateral triangle. This point, which describes one of the two intersection points of the chord with the circle, is kept fixed, while the second varies (so that the chord can move as a kind of pendulum). If we consider all possible points on the circle (relative to this second intersection point), we can see that the chord will rotate by a total angle of 180°, but that only those chords that intersect the arc subtended by an angle of 60° at the vertex (see Figure 1.2) will meet the requirement of being longer than the side of the inscribed equilateral triangle. Therefore, the probability P sought is given by the ratio:

$$P = \frac{60^\circ}{180^\circ} = \frac{1}{3}. \tag{1.1}$$

The *second solution* proposed by the French mathematician consists in first choosing an arbitrary direction and then considering all the chords that are parallel to that direction. Then, moving the

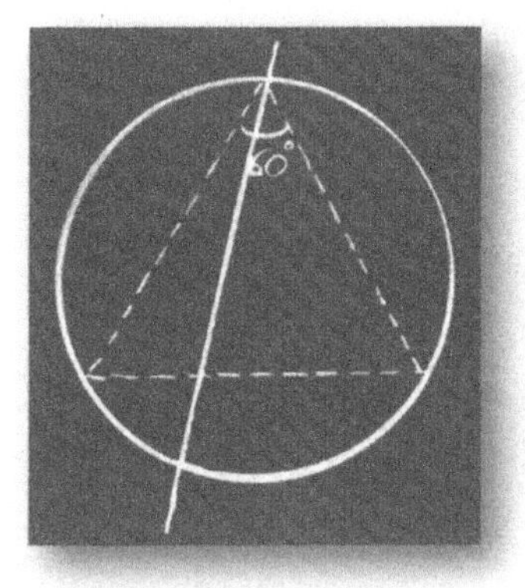

Fig. 1.2 A graphical representation of Bertrand's first solution.

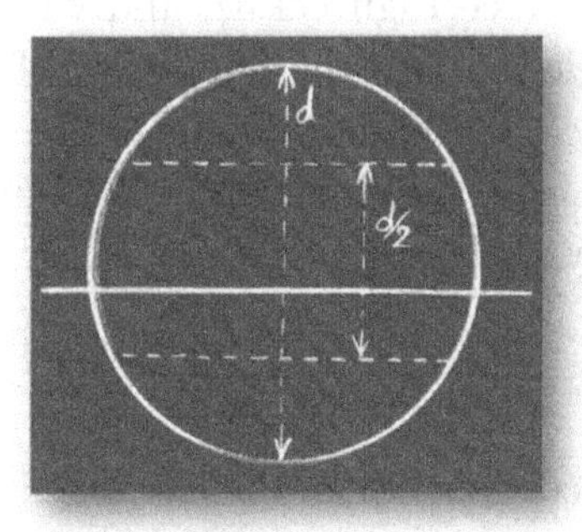

Fig. 1.3 A graphical representation of Bertrand's second solution.

chords along the circle, one observes that those intersecting its diameter (perpendicular to them) in its central segment, whose length is half the diameter of the circle (see Figure 1.3), satisfy the condition of being longer than the side of the inscribed equilateral triangle. This time, the probability is therefore given by the ratio:

$$P = \frac{\text{half diameter}}{\text{diameter}} = \frac{1}{2}. \tag{1.2}$$

Finally, the *third solution* proposed by Bertrand consists in choosing an arbitrary point inside the circle and consider it as the middle point of the chord. If we move this point around the entire area of the circle, we can observe that all the chords having their middle

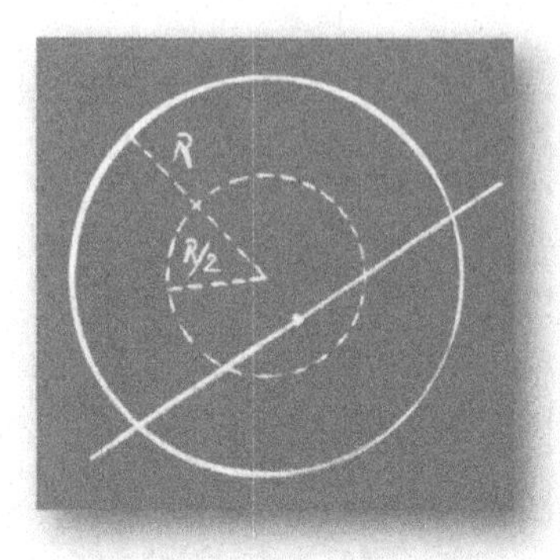

Fig. 1.4 A graphical representation of Bertrand's third solution.

point within an internal smaller circle whose radius is one half the radius of the big circle (see Figure 1.4), and which is the incircle of the equilateral triangle, satisfy the condition of being longer than the side of the inscribed equilateral triangle. The area of the internal circle being one fourth of the area of the big circle, this time, the probability is given by the ratio:

$$P = \frac{\text{area of small circle}}{\text{area of big circle}} = \frac{1}{4}. \tag{1.3}$$

In summary, proceeding with the above three arguments, all apparently valid, one gets three different values for the desired probability: $\frac{1}{3}$, $\frac{1}{2}$ and $\frac{1}{4}$.

What is the current status of Bertrand's paradox? Well, according to a recent analysis by philosopher *Nicholas Shackel* (Shackel, 2007), the situation is that after more than a century the paradox remains unsolved and continues to stand in refutation of the so-called *principle of indifference*, according to which: *alternative events for which there are equal reasons are equally likely to occur.*[1]

[1]The principle of indifference was originally formulated by *Jakob Bernoulli* as the *principle of insufficient reason*, and later on by *John Maynard Keynes*, who strenuously opposed the principle, and devoted an entire chapter of his book in an attempt to refute it (Keynes, 1921). The principle is usually assumed to incorporate a necessary truth about the relation between "possibilities" and "probabilities," i.e., that *possibilities of which we have equal (objective) ignorance have*

Even more recently, and more pessimistically, philosopher *Darrell P. Rowbottom* affirmed that Bertrand's proposed solutions to his question, which generate his chord paradox, are all inapplicable, so that there is no solace for the defenders of the principle of indifference, as it emerges that the paradox is much harder to solve than previously anticipated (Rowbottom, 2013).

This was the situation when we "stumbled" on what we believe is a compelling solution to the problem, which allows saving the validity of the principle of indifference. If we say that we "stumbled" on the solution it is because, while working on another famous theoretical problem namely the measurement problem in quantum mechanics, we realized that Bertrand's paradox was precisely one of the obstacles that stood between us and the solution to the problem. So by looking for a solution to the latter, we have also obtained, as a bonus, a solution to the former.

This intimate connection between the foundations of probability theory and the foundations of quantum theory, however, should not be a surprise. Indeed, it is not the result of a coincidence, since both these approaches to reality aim to describe systems that are subjected to specific actions, according to protocols that incorporate the presence of *irreducible fluctuations*, so much so that the results of these actions cannot be predicted with certainty, not even in principle (from which the necessary use of probabilities follows). In this sense, we could say that the founding fathers of probability theory, without knowing it, were actually quantum physicists *ante litteram*, as we will explain in more detail later in the text. We can also observe that the precise axiomatic formalization of probability theory, achieved by one of the most important and influential mathematicians of the twentieth century, the Russian *Andrey Nikolaevich Kolmogorov* (see Figure 1.5), took place exactly in the same years when the precise axiomatic formalization of quantum theory was

---

equal probabilities. It is generally assumed that its application suffices to solve probability problems and to find unique solutions to them. But this belief is precisely what has been undermined by Bertrand's three different "solutions," all three apparently based on such fundamental principle.

Fig. 1.5 A. N. Kolmogorov.

Fig. 1.6 J. von Neumann.

obtained by the Hungarian mathematician and physicist *John von Neumann* (see Figure 1.6). In fact, while Kolmogorov's founding text is dated 1933 (Kolmogorov, 1933), von Neumann's was published just a year before (von Neumann, 1932). This is to say that modern probability theory has developed in parallel with quantum theory, which in part also explains the difficulty, to quantum physicists, of understanding the true nature of the quantum probabilities, which are said to be *non-Kolmogorovian* precisely because they violate the famous classical axioms enunciated by Kolmogorov.

But let us go back to Bertrand's paradox. For didactical reasons, we will proceed contrariwise to the path that we have followed in our research, that is, we will use the solution of Bertrand's paradox to explain how it is possible to obtain a solution to the measurement problem.

The first thing we need to observe is that Bertrand's question actually hides two separate problems: an *easy* one and a *hard* one. These two problems must be carefully separated if one wants to get an unambiguous answer to his question.

The first problem, the easy one, is just about defining in a precise way the entity that is subjected to the statistical study, in our case the *chord* to be randomly drawn onto the circle. According to Bertrand, the chord in question is simply a *straight line*, that is, an *abstract entity* that is the mathematical modelization and idealization of different *possible material entities.* In fact, in practical terms, one can draw a line by throwing a sufficiently long straw onto a circle drawn on the floor, but also by tossing two pebbles, to be used then as the points to draw the straight line. But tossing a straw, or two pebbles, are two very different physical processes, which are not going to produce the same statistical distribution of straight-line segments.

Now, if we look closely at the logical construction of Bertrand's first solution, we can see that it describes a process where the line is constructed from two randomly chosen points. So, in its first solution, the French mathematician had in mind a physical process of the "tossing of two pebbles" kind. Instead, the second solution is undoubtedly a process of the "tossing of a straw" kind, even though it is a rather unusual tossing, since the orientation of the chord and its displacement towards the circle are determined in a sequential way by the thrower. Finally, its third solution is a mixed process, where a pebble is first tossed, to define the midpoint of the chord, and then a straw is tossed, to define its orientation.

While this ambiguity is maintained in Bertrand's statement, i.e., while in the question several *non-equivalent* questions remain mixed, referring to *different physical entities* (two pebbles, a straw, or a pebble and a straw), we will be dealing with an *artificially* created paradox, that is, a *false paradox.* In other words, to address Bertrand's paradox, the first thing to do is to make his question *operationally precise*, by specifying the nature of the material entity of which the chord is the abstract modelization.

For example, let us suppose that the entity is a straw (or rather, the idealization of a straw, that is, a very long, thin and stiff object). Would this allow us to determine that only Bertrand's first answer would be valid, and therefore its paradox would have suddenly vanished?

Things are not so simple. As we said, there are two problems in Bertrand's paradox. The easy problem is easy because it is just related to an ambiguity present in its statement, which allows for different non-equivalent modelizations of the entity subjected to the randomization. Once the ambiguity in the statement is removed, for instance by deciding that the chord is the abstraction of a straw-like entity, the easy problem is automatically solved. However, the hard problem remains: the one related to the *randomization process.*

Indeed, in his question Bertrand asks us to draw the chord onto the circle at *random.* But what does "at random" mean? Well, it means: *in no specific predetermined way.* In the case of a straw that is tossed towards a circle drawn on the floor, this means that we are in a situation of a *maximum lack of knowledge* about how the straw will in fact be thrown.

Of course, each throw will produce a specific outcome, in the sense that each throw will *select* a specific interaction between the thrower's hand and the straw, and once the interaction is selected, the trajectory of the straw will be perfectly determined, with the "hand–straw" interaction being one of a perfectly deterministic kind.

Now, considering that every interaction produces a perfectly deterministic outcome, we might be tempted to believe that the thrower, with her/his throw, selects a final chord on the circle. But this would be a mistake that is not without consequences. The thrower, in fact, does not select a final chord on the circle; rather, as its very name indicates, *the thrower selects a throw*! And this is where things become subtle, and interesting: *how does the thrower select a specific throw, i.e., a specific interaction between his hand and the straw?*

In order to select a specific throw, the thrower will first have to select *a specific way of throwing the straw.* Of course, we are not saying here that the thrower will do this in a conscious way, although she/he could certainly choose to do so. We are merely saying that at a more fundamental level, the thrower does not choose an interaction but *a way to choose an interaction.*

To try to explain what we exactly mean by this, a brief digression on the game of golf will be helpful. As is known, golfers have their own *style* of play, and in particular a very personal *swing*, that is, a special *way* of hitting the ball. For example, some players try to better control their shot by conferring a special *effect* to the ball. One possibility is to hit the ball so as to produce a clockwise rotation, which according to the *Magnus effect* will produce an in-flight trajectory curved to the right (if the curve is very marked, it is called a "slice;" if averagely marked, a "fade;" and if very slightly marked, a "push;" see Figure 1.7).

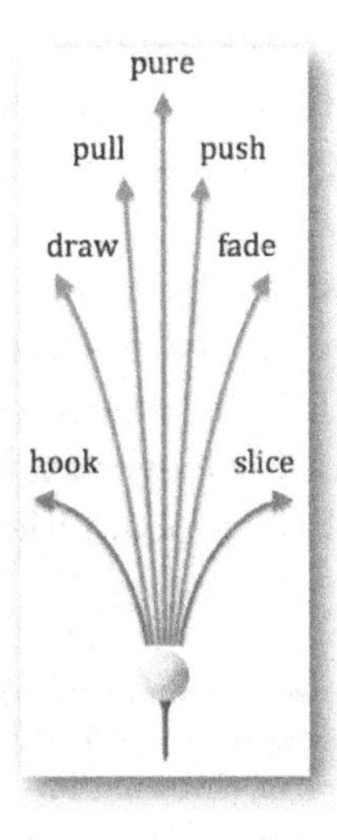

Fig. 1.7 The different effects with which you can hit a golf ball.

Similarly, if the style of the player is such as to produce a counterclockwise rotation of the ball, its trajectory will be curved to the left (giving rise to either a "hook," a "draw," or a "pull," depending on the degree of curvature of the trajectory; see Figure 1.7). Then there are players who use no effects at all, producing trajectories that are usually perfectly straight (called "pure"). Clearly, the style adopted by the players will take them more to the right-hand side or to the left-hand side of the golf course. Usually, players using the fade effect will tend to find themselves more often on the right-hand side of the course, while those using the draw effect will tend to wind up on the left-hand side of the course.

We can therefore say that these different styles of play will produce different statistics of impacts, as regards the *lateralization* of the ball onto the golf course. And it is clear that good golf players, even before choosing a trajectory for their shots, *will choose a way to hit the ball*, for example with a right–left effect, or a left–right effect.

In the case of throwing the straw onto the circle, the situation is the same, except that the choice about how to throw it will not be actualized voluntarily. But it is important to clarify that even if this choice were to be actualized voluntarily, this would not allow the thrower, in general,[2] to predetermine the final outcome of the throw. To choose a way of throwing does not mean to predetermine the specific interaction between the thrower's hand and the straw, but only to assign a specific *probability distribution* to the different possible interactions.

Now, it is possible to show, by means of an adequate modeling of the problem, that every way of throwing (i.e., every probability distribution associated with it) is able to produce a different probability that the obtained chord is longer than the side of the inscribed triangle. In other words, all the probabilities in the interval from 0 to

[2]If we say "in general," it is because we cannot rule out those very special throwing modalities that would allow throwers to perfectly control the trajectory of the straw, but these particular ways of throwing are only a very small subclass of all the possible ways of throwing, which is said to be of "zero measure" in the technical jargon, being entirely negligible.

Fig. 1.8 E. T. Jaynes.

1 are possible *a priori*, not only the three values historically indicated by Bertrand! Therefore, to solve the hard part of Bertrand's paradox means to explain how a single value can be obtained, when an infinite number of values are possible *a priori*.

To do this, we must follow the advice of the American physicist *Edwin Thompson Jaynes* (see Figure 1.8), who years ago proposed an interesting solution to Bertrand's paradox, using a *symmetry argument* (Jaynes, 1973). Our solution is perfectly compatible with that of Jaynes, but our approach addresses the problem from a perspective that in some ways is more fundamental. Jaynes rightly said that *when we solve a problem we should not use any information that is not given in the statement of the problem.* In our case, this means that we should not use any information relative to the way the straw is thrown. That is, we are in a state of ignorance not only about the outcome of the throw, but also, and especially, about the choice of the way of throwing! It is therefore in this condition of "double ignorance" that we must apply what *Nicholas Shackel* called the *principle of metaindifference.*

According to Shackel, just as the principle of indifference tells us that, in the absence of further information, we should give equal chances to equally possible events, in the same way, in the absence of further information, we should give equal chances to the different ways of producing random processes, that is, to the different ways of producing probabilities. In other words, since we do not know how the straw will be thrown, we have to give equal chances to all possible ways to throw it. And if we want to answer Bertrand's question, we must take the (uniform, i.e. arithmetic) average over all possible ways to throw a straw — an average to which we have given the name of *universal average.*

So, to solve the hard part of Bertrand's paradox, one has to calculate a universal average. The problem at this point is not only conceptual but also technical. However, it may also look like an unsolvable mathematical problem, as to take the arithmetic mean over all possible ways of throwing a straw means to average over all possible *probability distributions* ρ that characterize them. These, of course, are infinite in number, but that is not the problem: the problem is that it is an *uncountable infinity.*

More precisely, this means that the different probability distributions ρ (which technically speaking are called *generalized functions*) cannot be put in biunivocal (one-to-one) correspondence with the set of *natural numbers*, and of course, without the natural numbers, one cannot work out an arithmetic mean.

Let us explain this in more detail. Suppose that there are only two possible ways of throwing a straw, and that these two ways produce the probabilities P_1 and P_2, respectively, to obtain a chord that is longer than the side of the equilateral triangle inscribed in the circle. In this case, their average, let us call it $\langle P \rangle_2$, will simply be the sum of the two probabilities divided by 2, namely:

$$\langle P \rangle_2 = \frac{P_1 + P_2}{2}. \tag{1.4}$$

If, instead, the number of possible ways of throwing a straw is n, associated with the n probabilities $P_1, \ldots, P_{n-1}, P_n$, respectively, then, similarly, we have that their arithmetic mean $\langle P \rangle_n$ will be given by the weighted sum:

$$\langle P \rangle_n = \frac{P_1 + P_2 + \cdots + P_{n-1} + P_n}{n}. \tag{1.5}$$

Finally, if the number of possible ways of throwing a straw is infinite, but nevertheless *countable*, the average probability can always be calculated by considering the limit of the previous expression when the number n approaches infinity ($n \to \infty$).

But as we said, the different ways of throwing a straw are not countable, and if we cannot count them, neither can we calculate the average of the probabilities associated with them. Unless we find a way to make countable something that is not countable!

A typical example of an *uncountable* set of entities is the set of the *real numbers*, like those that describe the points that are located on any line segment. These points are not only infinite, but their infinity is of a higher order than that of the natural numbers.

However, each real number x can be described by a sequence of rational numbers r_n (called rational because formed by the ratio of two integers), such that when n tends to infinity, r_n tends to x. As an example, the (irrational) Euler's number e (used to define the *exponential function*) can be obtained as the limit of the rational numbers:

$$r_n = \frac{(n+1)^n}{n^n}, \tag{1.6}$$

when n tends to infinity ($n \to \infty$). The rational numbers, on the other hand, are perfectly countable. Thus, one can describe the entities belonging to an uncountable set by means of sequences of countable entities.

Can we do the same thing for the probability distributions ρ? The answer is affirmative. Indeed, it is possible to define sequences of special probability distributions ρ_n, called *cellular*, which can approximate with arbitrary precision any probability distribution ρ.

The distributions ρ_n are called cellular because they are formed by a finite number n of cells, and are such that in every cell they can take only two values: *zero*, or a specific *constant*, which is the same for all non-zero cells. In other words, the cellular probability distributions are functions whose definition domain is structured in two types of cells: the "empty" ones (if the distribution is zero) and the "full" ones (if the distribution is different from zero).

To use an analogy, a probability distribution is like an image of infinite resolution, while the cellular distributions of which they are

Fig. 1.9 Cellular probability distributions are similar to digital images: the higher the number of pixels, the higher the resolution of the image; in the ideal limit of an infinite number of pixels, the image perfectly reproduces the real subject, in every detail. Similarly, in the limit of an infinite number of cells, the cellular distributions are able to reproduce every possible probability distribution.

the approximations are like versions of the same image with a lower resolution, formed by a finite number n of pixels (see Figure 1.9).

More precisely, if a cell is empty, it means that the probability of observing the straw at the positions determined by the parameters contained in that cell is zero, whereas if it is full, it means that such probability is different from zero. And in the same way that appropriate sequences of rational numbers r_n can describe, in the limit $n \to \infty$, any real number x, one can also demonstrate that appropriate sequences of cellular distributions ρ_n can describe, in the limit $n \to \infty$, any possible probability distribution ρ.

The interesting thing is that the cellular probability distributions ρ_n are countable, and therefore so are the probabilities associated with them. This means that, to calculate a universal average, one can proceed in two steps. In the first step, one considers all the probabilities associated with the cellular distributions ρ_n formed by exactly n cells (their number grows exponentially with n, but for each n it is well defined), and then one calculates their arithmetic mean.

This is a calculation that can be made by *induction*, which is the equivalent in mathematics of the *domino effect*. Namely, we can proceed as follows: we calculate the average explicitly for some values, for example $n = 1$, $n = 2$, $n = 3$, etc. (which corresponds to manually bringing down the first domino tiles). Then, we have to guess the general formula for the average, valid for any value of n, and next we have to show that if it is valid for n, it is necessarily also valid for $n + 1$. In the metaphor of the domino, this is about demonstrating that if tile number n falls, then, inevitably, also the following tile has to fall. And since we made the first tiles fall by hand (by calculating explicitly the corresponding averages), we know that in the end they will all fall, which completes the proof by induction.

The second step, finally, is about identifying the limit $n \to \infty$ (n tends to infinity) of the average probability so obtained, which is an operation that does not present any particular problem. In this way, considering that every distribution ρ can be approximated with arbitrarily high precision by a sequence of cellular distribution ρ_n, when the number of cells is arbitrarily large (see Figure 1.10), by

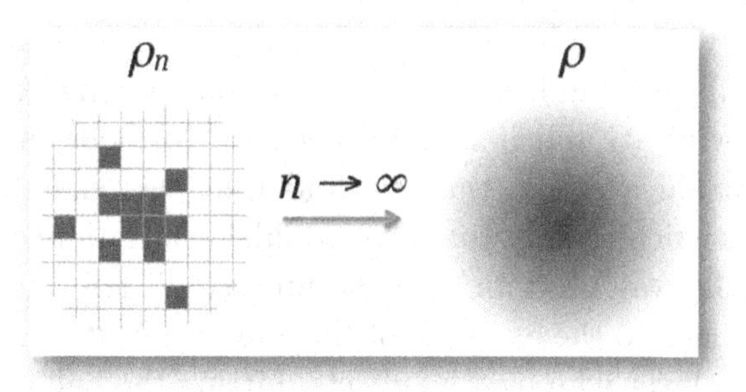

Fig. 1.10 In the $n \to \infty$ limit, the cellular distributions can describe any possible probability distribution.

calculating the average over all cellular distributions, then taking the limit $n \to \infty$, we can claim to have made a universal average, i.e., in our case, the average over all possible ways of throwing a straw. Working out in practice this calculation, one obtains the value:

$$\langle P \rangle_\infty = \frac{1}{2}. \tag{1.7}$$

Since this is the only possible value, Bertrand's paradox has thus been resolved!

Of course, the universal average may yield other values if instead of a straw we consider other physical entities (such as two pebbles, or a pebble and a straw) to produce the chord, but this, as we already explained, is due to the ambiguity that is present in the original formulation of Bertrand's problem, which needs to be disambiguated (easy problem) before proceeding to calculate the universal average (hard problem).

We conclude this chapter with an important observation. The abovementioned value obtained for the universal average is the same that we would have obtained if we had chosen a particular probability distribution, ρ_u, called *uniform*, as it attaches the same weight (the same probability) to every possible outcome of the throwing of the straw. In a sense, we could have expected this result from the beginning, as a uniform distribution undoubtedly expresses a condition of maximum ignorance about the outcome of a throw.

However, if we had used a uniform distribution from the beginning, we could not have claimed to be in a condition of maximum ignorance, as we would have known that the way the straw is thrown is of the "uniform" kind. But this is only one among an infinity of ways of throwing it, and it is only when we consider them all, and make a universal average, that we find that this exactly produces the same statistical result that we would have obtained by always throwing in the way described by the uniform distribution ρ_u.

To come back to the example of the golf player, it is important to distinguish a player who uses no effects, hitting the ball always along perfectly straight trajectories, from a player who uses all possible effects at random. The second player, on average, will keep the ball as straight as the first, in the sense that, on average, her/his trajectories will go equally to the left and to the right, but of course the playing mode of the first player will be very different from that of the second. The first player only explores a tiny portion of the golf course (its central part), while the second explores its full length and width. And the same is true for the "playing mode" subtended by a so-called universal average.

Chapter 2

The Measurement Problem

Now that we have explained how to solve the historic Bertrand's paradox, we are well equipped to face another famed and quite fundamental problem, namely the *measurement problem of quantum mechanics*. To understand what it is all about, we must first ask ourselves what it means to *measure*.

Usually, measuring is a *process of interrogative nature* by which we attribute to a given question an answer which corresponds to the outcome of the interrogative process. For example: *What is the circumference of the apple that we are holding in our hand at this moment*? To answer this question, we have to concretely measure the circumference of the apple and, to do that, we can use different equivalent instruments, such as a classic tailor's meter (see Figure 2.1).

In this case, we will hold the meter tight against the apple and roll it out along its circumference; then, making sure there is enough light, we will read the numerical value obtained. For example, if this value is 19 cm, the outcome of the measurement, that is, the answer to the above question, will be 19 cm.

Of course, although the measurement process allows us to know the circumference of the apple, we will all agree that its value existed even before it was measured, i.e., observed. In other words, measuring the circumference of the apple is a process that allows the observer to *discover* a property that is already *actual*, i.e., already existing, regardless of its measurement.

Fig. 2.1 Measuring the circumference of an apple by using a tailor's meter.

Before the advent of quantum mechanics, this latter statement would have been entirely self-evident, as it was clear to all that, apart from the inevitable experimental errors, measurement processes always gave *predetermined* (and in principle predeterminable) outcomes, as by definition the only thing they did was highlight properties of the entities subjected to the measurements that were *already actual* prior to the measurement. In short, the circumference of the apple is an *element of reality* that exists regardless of whether we measure it (observe it) or not.

This state of affairs changed dramatically when physicists began to measure the properties of *microscopic* entities, such as *electrons*, *protons*, *neutrons*, *atoms*, etc. In fact, we soon realized that the measurements of microscopic systems did not always yield predictable outcomes, even when the *states* of the entities subjected to the measurements were perfectly known. This meant that the interrogative process corresponding to a measurement was only going to determine a "spectrum of *possibilities*," corresponding to the different obtainable results, but that none of these possibilities could be considered to be already actual prior to its actual performance, with the measurement being precisely the process that was going to actualize one (and only one) among the different *potential properties*, *a priori* possible.

In other words, if in classical physics a measurement process was considered only to be a process of *discovery*, after the advent of quantum mechanics it was also (and especially) understood as a process of *creation*. If we compare the apple to a microscopic entity, it was as if the apple did not have an *a priori* circumference, and that it did not become real until the moment of its being measured.

This strangeness of the quantum measurements gave rise to several tentative explanations, associated with different possible *interpretations* of the theory. Basically, we can say that all these interpretations tried to do one and the same thing: explain the origin of the "*observer effect*" in the quantum measurements. Clearly, there are many other "strange" aspects of the theory which are considered to be poorly understood, and which also require an explanation, but the *measurement problem* unanimously remains a central one, and it is unthinkable to be able to explain the other peculiarities of the theory if we are not able to explain what it means, in the first place, to observe a microscopic system.

As we shall see, the notion of universal average will prove very important in proposing a solution to the measurement problem, in the sense that quantum measurements can be understood as *universal measurements*, i.e., as measurements that produce a *universal average*. But to understand what this means, we must proceed in stages.

We will start by explaining how a quantum measurement occurs in practice. To do so, we will use the example of one of the experiments that had the greatest impact on modern physics: the *Stern–Gerlach experiment*.

This particular experiment, which was carried out in Frankfurt by German physicists *Otto Stern* and *Walther Gerlach* (see Figure 2.2), in 1922, is a procedure by which we can measure a particular property of the microscopic entities: their *spin*. Actually, when Stern and Gerlach did their famous experiment, the spin had not yet been discovered, and it was only later that the Austrian physicist *Wolfgang*

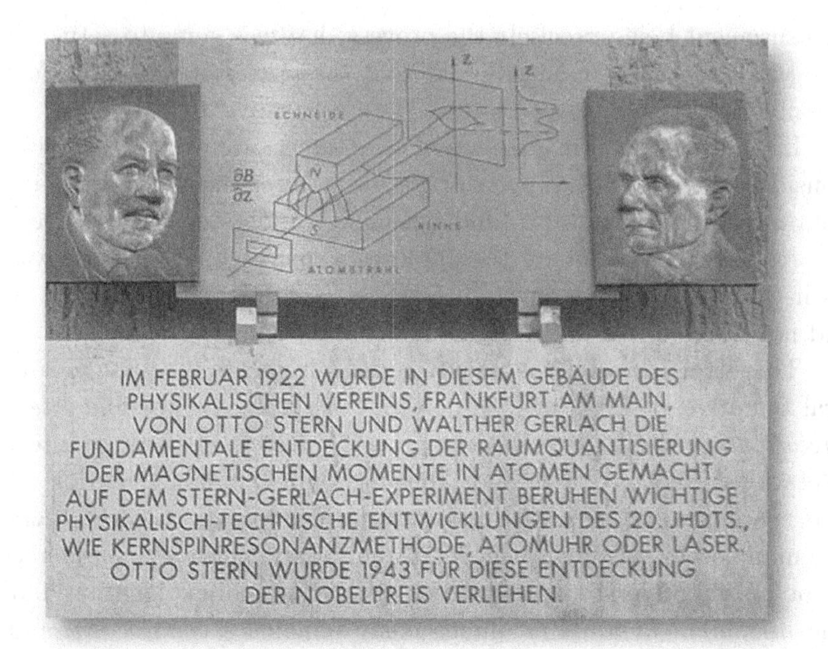

z
z
SCHNEIDE
$\frac{\partial B}{\partial z}$
N
S
ATOMSTRAHL
IM FEBRUAR 1922 WURDE IN DIESEM GEBÄUDE DES
PHYSIKALISCHEN VEREINS, FRANKFURT AM MAIN,
VON OTTO STERN UND WALTHER GERLACH DIE
FUNDAMENTALE ENTDECKUNG DER RAUMQUANTISIERUNG
DER MAGNETISCHEN MOMENTE IN ATOMEN GEMACHT.
AUF DEM STERN-GERLACH-EXPERIMENT BERUHEN WICHTIGE
PHYSIKALISCH-TECHNISCHE ENTWICKLUNGEN DES 20. JHDTS.,
WIE KERNSPINRESONANZMETHODE, ATOMUHR ODER LASER.
OTTO STERN WURDE 1943 FÜR DIESE ENTDECKUNG
DER NOBELPREIS VERLIEHEN.

Fig. 2.2 The commemorative plaque of the Stern–Gerlach experiment, located at the entrance of the former headquarters of the Institute of Physics, University of Frankfurt, Germany. The plaque reads: "In February 1922, the fundamental discovery of the space quantization of the magnetic moments in atoms was made in this building of the Physikalischer Verein, Frankfurt am Main, by Otto Stern and Walther Gerlach. The Stern–Gerlach experiment is the foundation of important physical and technical developments of the 20th century, such as the nuclear resonance method, the atomic clock, or the laser. Otto Stern was awarded the Noble Prize for this discovery in 1943."

Pauli hypothesized on the existence of this particular property of the microscopic entities, and that the results of their experiment could be fully understood, but this is another story.

What is a *spin*? Or rather: what do physicists usually think spin is? The typical statement is that a spin, for example the spin of an *electron*, is a kind of *intrinsic angular momentum*. If we think of the electron as a corpuscle, that is, as a sort of microscopic marble, the spin would then be the equivalent of a *rotational movement* of the marble around itself. Of course, the classic image of an electron as a micro-marble is deeply wrong, but it is undoubtedly the first one that

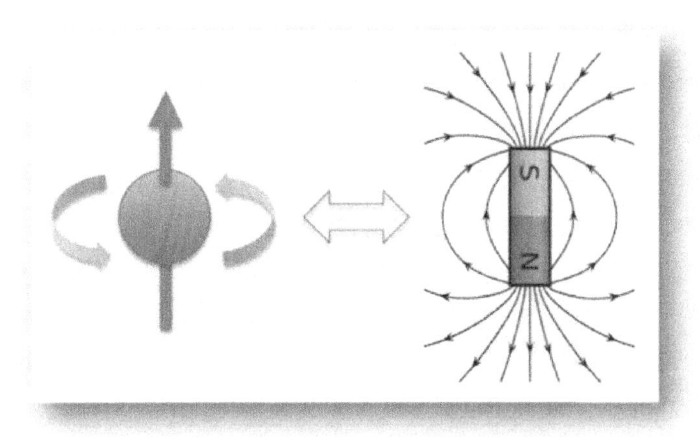

Fig. 2.3 An electrically charged corpuscle, revolving around itself, is the equivalent of a small magnet.

physicists have explored, also in the attempt to explain the nature of their spin.

Now, since an electron is an electrically charged entity, it is possible to associate a current loop with its movement of rotation around itself, and as is known, to a current loop corresponding to a *magnetic dipole*, that is, a *magnet* with a *north pole* and a *south pole* pointing in opposite directions. So, we can initially try to imagine the spin of an electron as a *micro-magnet*, carried by the electron itself, with a specific *spatial orientation* given by its north–south axis (see Figure 2.3).

The Stern–Gerlach experiment is a process that allows measuring the orientation of the spins of elementary entities, such as electrons, that is, the spatial orientation of the "small magnets" associated with them. In fact, the historical experiment performed by Stern and Gerlach was not performed on electrons, but on *silver atoms*, which contain electrons. On the other hand, all the electrons in an atom are paired, and when their number is odd, one of them remains inevitably solitary, as is the case in silver atoms. So considering that the spins of the paired electrons cancel out one another, we can say that in relation to its total spin, a silver atom is like a "big" and very massive electron.

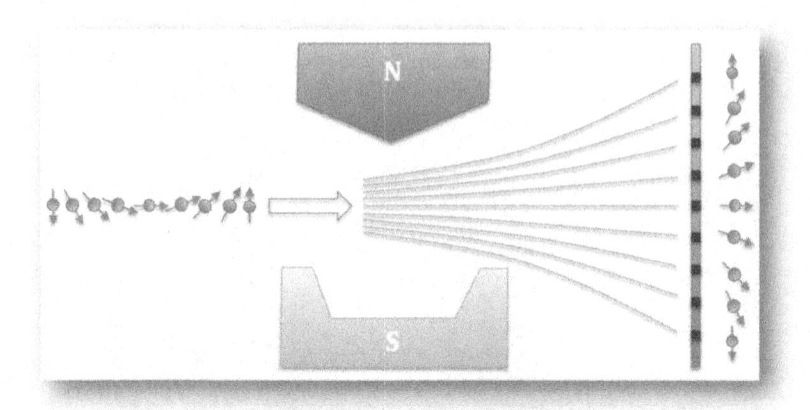

Fig. 2.4 A schematic representation of the Stern–Gerlach experiment, with the magnet generating the inhomogeneous magnetic field. If it is assumed that the electronic spins are like micro-magnets, then the experiment should produce a uniform distribution of impacts on the detection screen, as shown in the drawing, but this was not what Stern and Gerlach observed.

Thus, measuring the spin of a silver atom is roughly the same as measuring the spin of an electron, with the additional benefit that, because it is electrically neutral, a silver atom will not be subjected to the electromagnetic forces, such as the *Lorentz force*, which bends the trajectories of electrically charged particles when they pass through a magnetic field.

But let us come to the Stern–Gerlach experiment. Very simply, it consists in "firing" the silver atoms through a large magnet that generates an *inhomogeneous* (non-uniform) magnetic field, and then detecting the traces of their impacts on a glass plate placed at a certain distance from the magnet, oriented perpendicularly to the direction of the jet of atoms (see Figure 2.4).

If we think of a silver atom (more precisely, of his solitary electron) as a micro-magnet, then, since the magnetic field generated by the large magnet is not homogeneous, it will not attract (or repel) the south pole and the north pole of the micro-magnet with the same force. If for example the south pole of the micro-magnet points up, it will be attracted more strongly by the north pole of the magnet with respect to how its north pole will be attracted by the south pole of the magnet. This means that the micro-magnet will experience an upward force, which will accordingly deviate the trajectory of the atom, so that it will hit the detector screen in its upper part.

The opposite is true if the south pole of the micro-magnet points down. In fact, in this case it will be pushed with a weaker force by the south pole of the magnet with respect to how its north pole will be pushed by the north pole of the magnet. Therefore, this time the micro-magnet will be deflected downwards, and will hit the detector screen in its lower part.

What if instead the micro-magnet is oriented along the direction of its trajectory? In such a case, it will not experience any particular force, as the attractive and repulsive forces exerted by the magnet will cancel out each other, and the atom will end its run exactly at the center of the screen. And of course, for all the intermediate orientations of the electronic micro-magnets, they will experience intermediate deflections, producing impacts on the detector screen at intermediate positions.

In other words, if the micro-magnet model is correct, by firing a beam of silver atoms through an inhomogeneous magnet one would expect to observe a *uniform distribution of impacts* on the final detection screen, expressing the fact that in the atomic jet the micro-magnets are all oriented in completely arbitrary ways (see Figure 2.4).

But this was not what Stern and Gerlach observed! Indeed, instead of a uniform distribution of impacts, only those corresponding to *maximally deviated trajectories*, both upwards and downwards, appeared on the screen. And since, according to the classical model of the micro-magnet, each position on the detection screen is in correspondence with a specific orientation of the electronic spin crossing the magnet, the inevitable conclusion was that when the orientation of an electronic spin was measured along a given direction, only two values appeared to be possible: "up" or "down," that is, aligned along the direction of the magnetic field, or aligned in the opposite direction (see Figure 2.5).

Now, if we consider that the orientations of the spins that are shot through the magnet are completely random, the fact that only

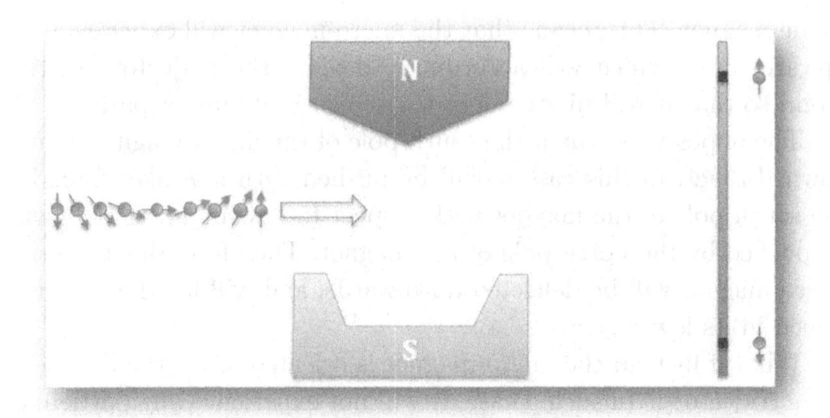

Fig. 2.5 Contrary to the hypothesis that the electrons' spins would be like small magnets, oriented along all possible spatial directions, Stern and Gerlach observed that only the impacts associated with the "up" or "down" orientation formed on the detection screen. In the drawing, the trajectories are no longer represented, as they are no longer *a priori* attributable to the silver atoms.

those with an upward, or downward orientation are observed is rather disconcerting. But let us try to better understand the dynamics of the experiment. Of course, the first reaction would be to think that if only the orientations "up" and "down" are observed, this is because only silver atoms having their electronic spins oriented in those two specific directions actually go through the magnet. This assumption, however, can easily be discarded. Indeed, it is always possible to use a magnet similar to that of the experiment to filter the unwanted orientations, and obtain silver atoms whose spins point along predetermined directions.

This operation of initial filtering is called *preparation* in physics. So, instead of shooting into the magnet atoms whose spins are oriented in an arbitrary manner, it is possible to prepare them so that their orientation is determined in advance. More exactly, a physicist will say that the *state* of the spin (of the valence electron contained in the silver atom) is perfectly known. And if the orientation of the spin is known, it is expected that the atom will leave a trace on the detection screen exactly at the position corresponding to the orientation in question.

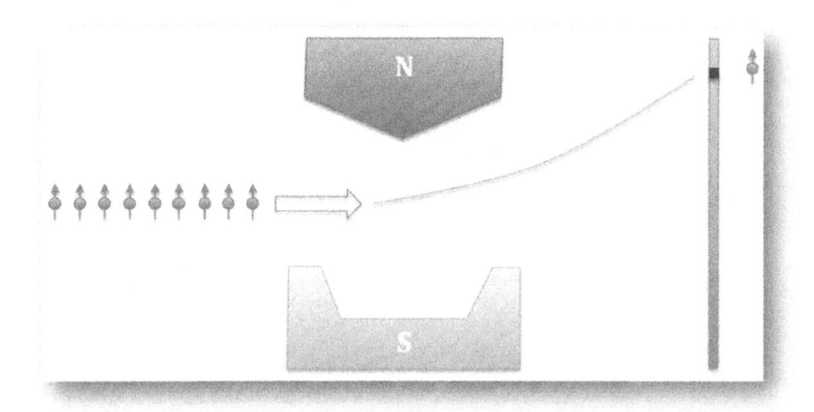

Fig. 2.6 If the spins are all prepared in the "up" state, the impacts on the detection screen will all correspond to a maximum upward deviation of the (alleged) trajectories.

In fact, the Stern–Gerlach experiment is meant to measure the orientation of the spins, and if the orientation is known in advance, it is natural to expect that the experiment will be able to reveal exactly such orientation, that is, *to confirm it.* We are saying something that apparently is perfectly self-evident. Yet, things do not work exactly this way.

Let us suppose that the silver atoms are prepared in such a way that all their spins point upwards ("up" state), that is, along the south–north direction of the magnet. In that case, consistently with our classical expectation, we will observe on the screen impacts corresponding only to a maximum upward deviation of the trajectories (see Figure 2.6).

Similarly, if we suppose that the silver atoms are prepared in such a way that all their spins point downwards ("down" state), that is along the north–south direction of the magnet, then, in accordance with our classical expectation, we will observe only impacts corresponding to a maximum downward deviation of the trajectories (see Figure 2.7).

What happens if instead we send through the magnet silver atoms prepared in such a way that their spins are neither oriented along

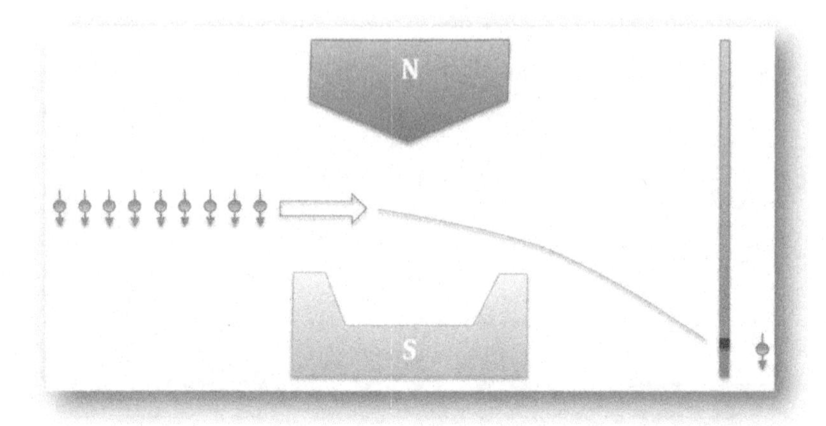

Fig. 2.7 If the spins are all prepared in the "down" state, the impacts on the detection screen will all correspond to a maximum downward deviation of the (alleged) trajectories.

the "up" direction, nor along the "down" direction? As we have seen, according to the classical expectation, that is, according to the idea (or rather, the preconceived idea) that a spin is a kind of micro-magnet, all the possible impacts should be observable, except for those corresponding to the maximum upward and downward deviations (see Figure 2.4). But this, as we have said, was not what Stern and Gerlach observed (see Figure 2.5).

So, when an electron's spin is in such a state that its orientation is neither *parallel* nor *antiparallel* to the magnetic field, this will nevertheless produce an impact corresponding to either a maximum upward deviation or a maximum downward deviation. And since the spin initially did not possess such orientations (relative to the direction of the magnetic field), we must conclude that it is the very measuring process that *created* them!

In summary, if the electron's spin is initially "up," we can predict *with certainty* that the measurement's outcome will be "up." Similarly, if the electron's spin is initially "down," we can predict *with certainty* that the measurement's outcome will be "down." But if the electron's spin is not initially "up" or "down" (i.e., it points in any other direction), *we have no way to predict with certainty what the*

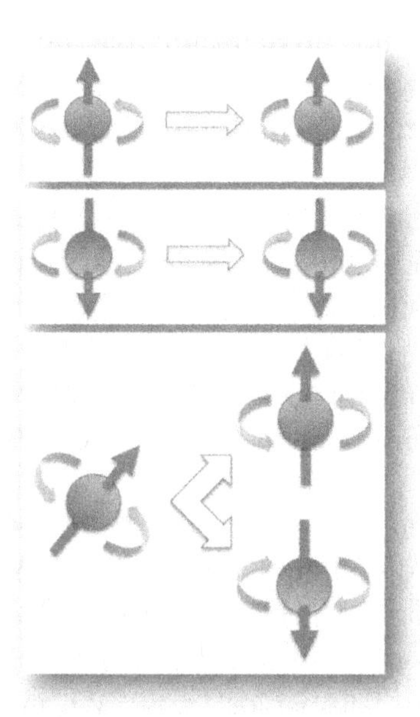

Fig. 2.8 If the preparation is "up," the measurement will invariably yield "up;" if the preparation is "down," the measurement will invariably yield "down;" if the preparation is neither "up" nor "down," the outcome of the measurement cannot be predetermined, but can only yield "up" or "down."

outcome of the measurement will be. Also, it can give the outcome "up" or the outcome "down," but no other outcome (orientation) will ever be observed (see Figure 2.8).

Therefore, the situation where the initial spin is oriented along the magnetic field involves a measurement process that is not only *deterministic*, in the sense that its outcome is predetermined, but also *non-invasive*, in the sense that the measured orientation corresponds exactly to that originally possessed by the spin, and in this sense the measurement process *does not change* the state of the spin.[1]

On the other hand, the situation in which the incoming spin is *not* oriented along the magnetic field involves a measurement process that is not only *indeterministic*, in the sense that its outcome is

[1]Of course, the absorption of an atom by the detector is a highly invasive process, but only the spin is considered in our discussion.

neither predetermined nor predeterminable, but also invasive, in the sense that the measured orientation does not correspond to the one initially possessed by the spin, and thus the measurement *does change* the state of the spin.

At this point, a natural question arises. If it is true that when the initial orientation of the spin is arbitrary we cannot predict the outcome of the measurement, which will be either "up" or "down" (and nothing else), *how does the selection of the final state occurs*? Is it completely random, or does it depend in some way on the initial orientation of the spin? And if it does, how can we express such dependency?

To answer this question, it is necessary to perform many experiments and see in detail what happens. More precisely, we have to choose a direction for the initial spin of the silver atom that is shot through the magnet, then write down what is the impact we have obtained on the screen: upper or lower. We must then repeat the experiment, always with the same initial orientation, and note once again the outcome, and so on, for a number N of times.

Suppose that N_{up} is the total number of impacts obtained in the "up" position of the screen, and N_{down} the total number of impacts obtained in the "down" position of the screen. If N is very large, we can deduce the probability P_{up} that the outcome is "up," for that particular initial orientation, by simply calculating the ratio:

$$P_{up} \approx \frac{N_{up}}{N}. \tag{2.1}$$

Similarly, it is possible to obtain the probability P_{down} that the outcome is "down," always for that particular initial orientation, by calculating the ratio:

$$P_{down} \approx \frac{N_{down}}{N}. \tag{2.2}$$

Of course, since $N_{up} + N_{down} = N$ (only the outcomes "up" and "down" are possible), the two probabilities will necessarily add up to one, that is: $P_{up} + P_{down} = 1$.

After having performed N experiments (that is, N measurements) with a particular initial orientation, we must perform N more, using a different initial orientation, and then another N number of experiments, with yet another initial orientation, and so on. Analyzing all the probabilities obtained, we can then try to figure out if there is a general *rule of correspondence* between the initial state of the incoming spin (its spatial orientation) and the probability of observing the final state (i.e., the outcome) "up" or the final state "down."

In doing so, it is not difficult to see that the important parameter in determining the probabilities is the angle θ between the initial orientation of the spin and the south–north direction of the magnetic field (which corresponds to the direction of the "up" outcome state; see Figure 2.9), and that the probabilities in question can be fully expressed as a function of it using the simple trigonometric function of the *cosine*[2]:

$$P_{up}(\theta) = \frac{1+\cos\theta}{2}, \quad P_{down}(\theta) = \frac{1-\cos\theta}{2}. \tag{2.3}$$

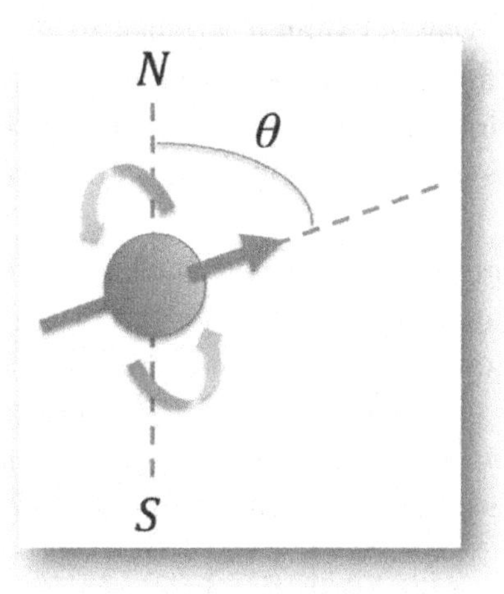

Fig. 2.9 The angle θ between the initial direction of the spin and the south–north direction of the magnetic field.

---

[2]In a *right triangle*, the *cosine* of an angle θ is defined as the ratio between the length of the *cathetus* adjacent to θ and the length of the *hypotenuse*.

This rule, which allows one to deduce the probabilities of the different outcomes of the measurement as a function of the initial and final states of the measured entity, is known in quantum physics by the name of *Born rule.*

We can observe that when the incoming spin points towards the north, the angle θ is zero, and since the cosine of $0°$ is 1, the Born rule tells us that in this case $P_{up} = 1$, i.e., that the outcome "up" is certain, consistent with what we have previously observed.

Similarly, when the incoming spin points towards the south, the angle is $180°$, and since the cosine of $180°$ gives -1, we obtain in this case $P_{down} = 1$. In other words, the rule covers all possible situations, those in which the outcome is predetermined, for the values $\theta = 0$ and $\theta = 180°$, and those for which the outcome is not predetermined, corresponding to the values of the angle θ that are other than zero and one hundred and eighty degrees.

Of course, the Born rule is expressed here in the particular situation of the measurement of the spin of an electron (the valence electron of the silver atom), but it is possible to formulate it in a much more general way, so that it applies to any kind of physical entity subjected to any kind of measurement. But for our discussion, this simplified version of the rule will be more than enough, since it already embodies all the mystery of the so-called *measurement problem.*

But what would be this measurement problem? Well, simply, it is about explaining the nature and structure of this very rule. Indeed, the rule allows us to calculate the different probabilities in a very precise and effective way, but the question is: *What is the origin of these quantum probabilities*? And why are we dealing with probabilities when we measure a physical quantity associated with a quantum entity?

Typically, probabilities are useful tools for quantifying a situation of *lack of knowledge.* Thus: *To what kind of lack of knowledge are the quantum probabilities referring to*? What exactly do we not know when we perform a quantum measurement, so as to require the use of probabilities to describe its possible outcomes (with the exception of particular initial states, such as when the spin is parallel or antiparallel to the magnetic field)?

These questions have haunted physicists since the early days of quantum theory. Indeed, the theory does not explain the origin of the Born rule, which is not derived from any underlying principles but simply postulated. Furthermore, since the rule determines what are the possible outcomes of whatever measurement process of a physical entity, our understanding of the very nature of the observed entities, and of the observational processes in general, also depends on it.

Now, if we assume that quantum mechanics is a complete theory, and that each measurement process (that is, each process of observation of a physical quantity) is universally described by the Born rule, we immediately face an insidious problem, summarized by the question: *Who measures the measuring instrument*? Indeed, the measuring instrument is also a physical system, which by hypothesis is subjected to the quantum laws. So, also the reading of the instrument by the human experimenter will be the equivalent of a measurement process, subjected to the Born rule, thus describable only in terms of probabilities. Why then, when we humans observe the impacts on a detection screen, for example in a Stern–Gerlach experiment, do we not observe potential happenings, but actual happenings?

We humans, if we consider our physical body, are also entities that by hypothesis obey the laws of quantum physics. So why is it that, when using our eyes and the processing capacity of our brain to observe a detector screen, we can see specific impacts? Why does the Born rule not apply as such, in our case? Why do we not observe "probabilities of impacts" (whatever that means), but very concrete and well-defined impacts? *How does this passage from probabilities to actualities occur*?

If we are not willing to abandon the idea that quantum mechanics (in its standard formulation) is a complete theory, then there are typically two possible ways out of this impasse. The first is to assume the existence of "extra-physical" entities, not subjected to the quantum laws, able to do what the physical systems are not able to do: transform abstract probabilities in concrete actualities.

As much as this hypothesis may seem speculative, and decidedly rather implausible, it was nonetheless endorsed by physicists such as *John von Neumann* (von Neumann, 1932) (already mentioned above), *Fritz London*, *Edmond Bauer* (London & Bauer, 1939), *Eugene Wigner* (Wigner, 1961), and more recently by *Henry Stapp* (Stapp, 2011), just to mention some of the best-known names. According to these authors, the extraphysical entities responsible for the processes of actualization of the quantum potentialities are none other than the very consciousnesses of the experimenters, understood here as "abstract realities" not entirely reducible to the activity of their brains.

However, this interpretation of the quantum measurement processes, sometimes called the *von Neumann–Wigner interpretation*, presents a number of difficulties. For example, if it were true that any conscious observer is able to attribute, individually and in a non-predetermined way, a specific value to the observed physical quantity, how is it possible that different observers always agree in establishing what is the final outcome of an experiment?

But the weakest part of this "psychophysical" explanation is that actually it explains nothing, as it simply *shifts the problem*: if before the problem was about explaining what happens in the interaction between a measuring instrument and a measured system, now the problem is to explain what happens in the interaction between the measuring instrument and the non-material consciousness.

The second way out of the impasse of the measurement problem, without giving up the hypothesis of completeness of quantum theory, is by quite simply eliminating the problem at its root. This is what the American physicist *Hugh Everett III* (see Figure 2.10) proposed, in the 1950s, in what is today known as the *many-worlds interpretation* (DeWitt & Graham, 1973; Everett, 1957).

According to this very radical view, the solution of the dilemma of quantum indeterminism lies in the acceptance of the fact that we live in a world where every possible outcome of an experiment is always actualized, but in different *parallel universes* (see Figure 2.11). In other words, according to Everett, our reality would be nothing

Fig. 2.10 H. Everett III.

more than a multiverse (or multiworld) partitioned into an infinity of constantly growing parallel universes, where every possibility is in fact always realized.

The many-worlds interpretation has gained considerable appeal in recent years, particularly among cosmologists. However, as is the case of the von Neumann–Wigner interpretation, it does not so much solve the problem as to shift it. In fact, even if we accept the reality of parallel universes, what Everett's theory is unable to explain is how we get to the quantum probabilities, i.e., it leaves the question unanswered of why we would obtain exactly those probabilities, with those specific values, rather than any other probabilities, with other values.

This difficulty is common to the interpretation of von Neumann–Wigner and that of Everett. They both put in place new entities to explain the transition from the probabilities to the actualities: in the former, it is the consciousness that produces the transition, while in the latter, it is reality itself that automatically actualizes any possibility, in a specific universe. But in both cases, the Born rule is not really explained, in the sense that they identify no credible

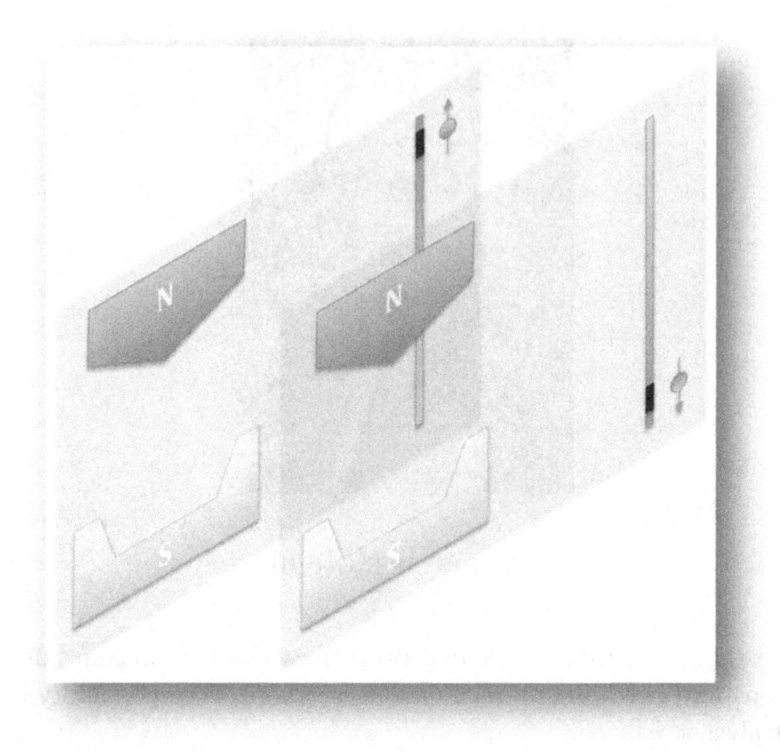

Fig. 2.11 According to the many-worlds interpretation, when a silver atom passes through a magnet of the Stern–Gerlach kind, it will be observed in the "up" part of the screen in one universe, and in the "down" part of the screen in another universe, not communicating with the first.

mechanism (be it physical or psychophysical) which would allow to derive it (that is, to deduce it) according to a non-circular logic.

Of course, it would be an extreme oversimplification to say that these are the only two approaches to solving the measurement problem, if we are to preserve the completeness of quantum theory. In fact, there are many other approaches to quantum physics that attempt to explain the nature of the observational processes, and it is not always easy to determine whether they provide only a different interpretation of the objects of the theory (with respect to the standard interpretation), or represent a new theoretical model that is also able to provide different predictions in certain experimental contexts.

The purpose of this booklet is not, however, to review all the interpretations that have been proposed over the years, but to explain how we arrived at a solution of the measurement problem that on the one hand takes the standard formalism of quantum physics very seriously, and on the other completes it, thus making it possible to derive the Born rule and explain what happens "behind the scenes" of a quantum measurement process. As we shall see, the notion of *universal average*, which we used to solve Bertrand's paradox, will play a key role in the solution.

Chapter 3

The Hidden-Measurements Interpretation

To explain how we can build a model that is able to complete the formalism of quantum theory by adding a detailed description of the measurement process, we will continue exploring the example of the Stern–Gerlach experiment, which is the simplest possible prototype of a quantum measurement. Later, we will explain how the model can be generalized to describe more complex experimental situations.

Let us suppose that the spin of the valence electron, in the silver atom, has been prepared in a specific *initial state*, before being fired through the magnet. In quantum physics, the state of a physical entity is described by a particular mathematical object: a *vector*, which belongs to a particular *vector space*, called the *Hilbert space*. In the case of the electronic spin, each state can be described, in a univocal way, as a *direction vector of the three-dimensional Euclidean space*. This direction is such that, if the south–north axis of the Stern–Gerlach magnet was oriented along it, the experiment would give the "up" outcome with certainty (see Figure 3.1).

We will use the Greek letter ψ to denote this state and we will represent it as a simple point on the surface of a sphere of unit radius (called the *Bloch sphere*, as it was the Swiss physicist *Felix Bloch* (see Figure 3.2) who in 1946 had the idea of representing the spin states of entities such as electrons in this way). We can visualize this point as a *material point*, that is, as a *point-like* corpuscle positioned exactly at that particular location on the surface of the Bloch sphere. This is

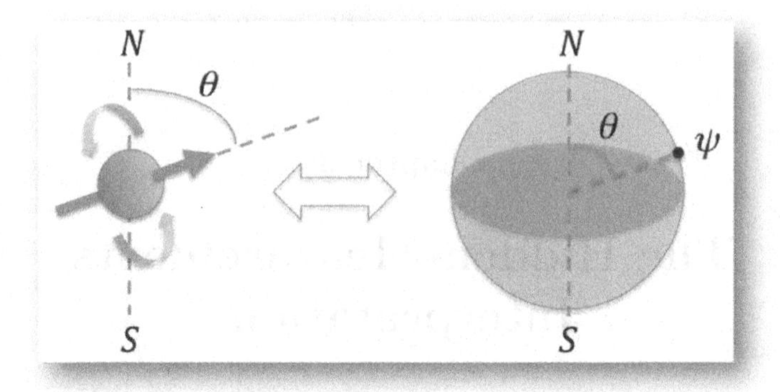

Fig. 3.1 The state of an electronic spin can be represented as a material point on the surface of the Bloch sphere. Since the Bloch sphere is of unit radius, only two angles are required to describe a point on its surface: the polar angle and the azimuthal angle. In the drawing, only the polar angle θ is represented.

Fig. 3.2 F. Bloch.

not to assert, however, that the spin of an electron is really a material particle on a sphere; rather, we can simply visualize it (i.e., modelize it) as if it were so. In other words, it is a structural analogy. The particle in question is obviously an abstract entity, representative of a state in an abstract state space, but it will allow us to better explain our reasoning if we *pretend* it were a real material corpuscle.

As we have seen, when the spin is prepared in the state ψ, and subjected to a measurement of its orientation through a Stern–Gerlach apparatus, only two outcomes are possible: "up" and "down." The outcome "up" corresponds to the spin pointing to the north pole of the magnet, while the outcome "down" corresponds to the spin pointing to the south pole of the magnet. We denote by ψ_{up} and ψ_{down} the spin states corresponding to these two outcomes. In the Bloch sphere, they correspond to a material particle located at its geographic north and south poles, respectively.

To explain the quantum measurement process, we must be able to identify a mechanism that can move the corpuscle from the position ψ to the position ψ_{up} or ψ_{down}, such that the final outcome is not predeterminable and that the probabilities of the two outcomes correspond exactly to those predicted by the Born rule (see Figure 3.3).

As we explained, these probabilities depend on the polar angle θ between the initial orientation of the spin and the measurement direction of the magnet. Therefore, the mechanism in question will have to account for the fact that the outcome of the measurement statistically depends on the initial state, even though it is unpredictable (provided that θ is different from $0°$ and $180°$).

This mechanism, to tell the truth, will have to be able to do much more: it will also have to be generalizable to more complex experimental situations, for example situations in which the number of possible outcomes is N, with N being arbitrarily large, and even situations where some of these outcomes may not be distinguishable by the experimenter. But let us go step-by-step and begin by explaining what this mechanism is in the situation of the measurement of the spin of an electron, which is a measurement only involving two possible outcomes ($N = 2$).

To do this, it is important to observe, as we already noted, that in the Bloch sphere the physical states correspond to the points *on the surface* of the sphere. The points inside the sphere, instead, are typically used to describe the so-called *mixed states*: conditions under

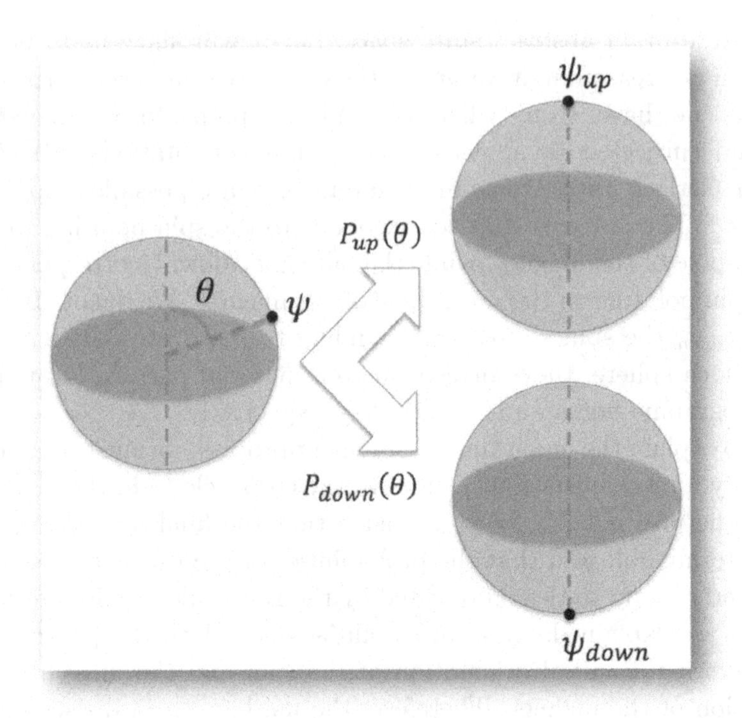

Fig. 3.3 A schematic representation of the measurement process of an electron's spin in the Bloch sphere. The corpuscle that represents the spin state passes from the initial condition ψ to the final condition ψ_{up}, with probability P_{up}, or to the final condition ψ_{down}, with probability P_{down}.

which the specific state of the entity in question would not be fully known to the experimenter. The point at the center of the sphere, for example, would correspond to a situation in which the experimenter has no knowledge of the state at all, while the points between the center and the surface represent increasingly specific knowledge of the state, with the points exactly on the surface being associated with the so-called (non-mixed) *pure states*, corresponding to a situation in which the experimenter has full knowledge of the state.

On the other hand, if it is true that the points inside the sphere can be used to describe these particular conditions of *subjective ignorance*, it is also true that the mathematical formalism of quantum theory remains somewhat ambiguous in this regard. In fact, one and the same point inside the sphere is able to represent an infinite number of mixtures of different states, that is, of different subjective conditions of ignorance, and this suggests that the mathematical entities describing the mixed states may also play a role in the description of pure states of a different kind than those usually represented by the points at the surface of the sphere.

Now, as we explained, to derive the Born rule and solve the measurement problem, it is necessary to complete the quantum formalism, whose standard version does not allow to describe what happens, specifically, during a measurement process. A natural way to do this, which we will adopt, is to assume that the points inside the Bloch sphere are also able to label pure states (i.e., situations of maximum knowledge), and that these are precisely those states that are missing in the standard formalism and are needed to explain the observational processes.

In other words, our hypothesis is that in the course of a measurement process the material particle representative of the state of the spin entity will penetrate the sphere, and that it will do so in a very particular way, which we will now explain. It is a movement generally happening in *three stages*, but in the simple case of the measurement of an electron's spin, only two of these stages will be present.[1]

To describe the process, we have to represent within the Bloch sphere not only the point representative of the spin state, but also the action of the measuring apparatus. We will call this particular description the *extended Bloch representation*, where the term "extended" refers to the fact that the measuring instrument is also represented.

To do this, we first observe that a measuring instrument is always relative to a *physical quantity* (usually called *observable* in quantum physics), namely the physical quantity that the instrument allows to observe, that is, to measure. In the present case, it is the physical

[1]The third stage is relevant only in the so-called *degenerate measurements*, where not all the outcomes of an experiment are distinguishable. In order not to unnecessarily complicate our discussion, however, we will not be concerned with such measurements here.

quantity corresponding to the value of the spin along the direction of the magnetic field. As we have seen, only two outcomes are possible, corresponding to the states ψ_{up} and ψ_{down} of the electron's spin. These two states are associated with the north pole and the south pole of the sphere, respectively (see Figures 3.1 and 3.3).

To represent the action of the measuring instrument, we will consider, in addition to these two antipodal points, all the intermediary points passing through the center of the sphere. To this end, we can imagine having stretched an *elastic band* between the two points, firmly anchored to them, and which we assume possesses the following three special properties.

First, the elastic band is *attractive*, in the sense that it will exert a strong attractive force onto the particle as soon as it is applied to the sphere, causing the particle to plunge into the sphere, along a rectilinear "falling" trajectory, *orthogonal* to the direction of the elastic.

This immersive (or "falling") movement will stop as soon as the corpuscle reaches the elastic (see the first three drawings of Figure 3.4), and when this happens, the corpuscle is no longer able to move (the attraction exerted by the elastic being maximum), so that it remains firmly anchored (stuck) to the elastic band, at the exact point where the contact took place.

The second property of the elastic is that it is *unstable*, in the sense that sooner or later it will disintegrate. In other words, and to put it in a simple way, the elastic is *breakable*, and therefore at some moment it will break, at an unspecified point. This breaking process is by definition completely *indeterministic*, and does not depend in any way on the initial state ψ of the corpuscle; rather, it depends solely on the nature of the elastic.

The third property of the elastic, as its name indicates, is that it is *elastic*. Therefore, since its end points are anchored at ψ_{up} and ψ_{down}, as soon as it breaks, the two resulting fragments will contract towards their respective anchor points. As a consequence of this contracting movement (collapse), the corpuscle representative of the state of the electron's spin, being attached to one of the two

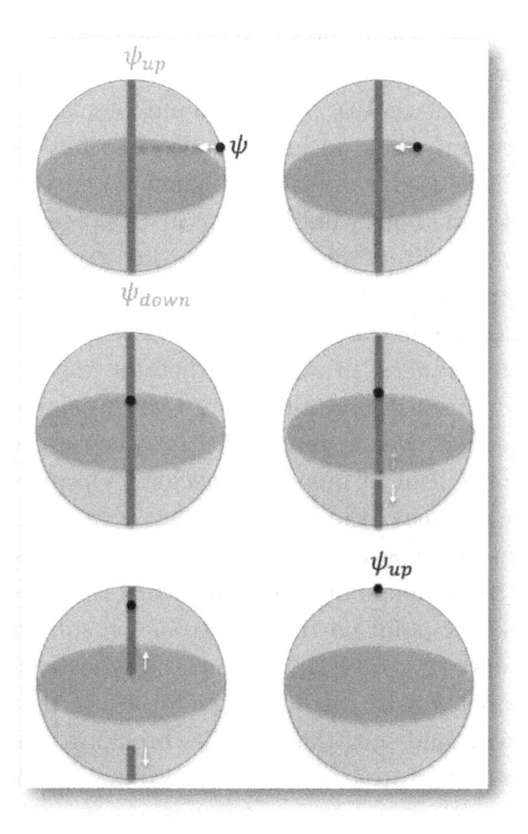

Fig. 3.4 A representation of the measurement process of an electron's spin. In the first, purely deterministic stage of the process, the particle representative of the spin state plunges into the sphere, sticking to the elastic band, stretched along the south–north axis. Then, in the second stage, the elastic breaks, at some unpredictable point, bringing the corpuscle again to the surface of the sphere at one of the two anchor points — in this case the one corresponding to the "up" state.

fragments, will be carried along and reach one of the two anchor points, ψ_{up} or ψ_{down}. This will correspond to the final state of the spin, after the measurement has been completed (see the last three drawings of Figure 3.4).

Which one of these two points will be reached by the corpuscle: ψ_{up} or ψ_{down}? This of course will depend on the breaking point of the elastic. If it is located between the position of the particle on the elastic and ψ_{down}, as in the situation described in Figure 3.4, the final outcome will be ψ_{up}; if it is located between the position of the particle on the elastic and ψ_{up}, the final outcome will be ψ_{down}.

Let us reflect on what we have obtained so far. In constructing the model, we have used everything we knew about the nature of the electron's spins, and about what happens in an experiment measuring them, as observed by Stern and Gerlach. More precisely, we have used the fact that spins can be represented as points on the surface of the unit Bloch sphere, and that, following a measurement process, the initial spin state changes, in a non-predictable way (barring exceptional initial conditions), but that only two outcomes are possible. These two outcomes, being *pure states* (as they are understood in the standard formalism of quantum mechanics), are also described as points on the surface of the sphere, so that a measurement process, viewed from the perspective of the Bloch sphere, is a process that allows to move from one point on the surface of the sphere (the initial state ψ), to another point on the surface of the sphere, chosen from two possible points (ψ_{up} or ψ_{down}). Since the process is indeterministic, we have modeled it by introducing a structure that breaks in an unpredictable way, and such that after its breaking it is able to bring the corpuscle representative of the state into one of the two possible final positions.

Of course, in place of an elastic we could have imagined other unstable structures able to promote a similar indeterministic process. On the other hand, there is a reason why we have chosen a one-dimensional structure of this kind, in addition to its simplicity. Only this kind of structure, as we will see later, can be naturally generalized when more complex experimental situations are considered.

We can observe another important aspect of the model of the elastic. If the initial position of the corpuscle is ψ_{up} or ψ_{down}, that is, in correspondence with one of the two anchor points of the elastic, there will be no initial movement of immersion into the sphere because it is already in contact with the elastic. Moreover, whatever will be the breaking point of the elastic, it will not have any repercussion on the corpuscle, which will remain exactly where it is located, in its initial position.

This means the operation of the elastic is compatible with what we have previously seen, namely that if the spin is initially in the state "up," the measurement will give with certainty "up" (see Figure 2.6), while if the spin is initially in the state "down," it will give with certainty "down" (see Figure 2.7). In quantum physics, the states whose measurement give an outcome that is certain in advance are called *eigenstates* of the physical quantity in question (which in our case is the value of the spin along the magnetic field direction).

We can thus observe that one of the properties of quantum measurements is that they produce a "collapse" of the state of the measured entity to one of the possible eigenstates of the observable in question. In other words, the results of a measurement are its eigenstates, that is, those states that will be the same with certainty if the measurement is repeated (von Neumann called the measurements having this ideal property *measurements of the first kind*, to distinguish them from those of the second kind, where the final state is not necessarily an eigenstate).

At this point, we can say that we have used everything we knew about the measurement of an electron's spin. We went a bit further compared to the standard formulation of quantum theory, by considering the points within the Bloch sphere as also representatives of possible pure states that a spin entity is able to assume during a measurement. And we have represented the indeterministic mechanism of the measurement by means of a one-dimensional unstable structure, placed inside the sphere, representative of the *potentiality region* that produces the passage from the abstract probabilities to the concrete actualities.

What *apparently* our modeling is still lacking for it to be considered a solution to the measurement problem (for the time being only for an electron's spin measurement), is the ability to quantitatively predict the exact value of the quantum probabilities $P_{up}(\theta)$ and $P_{down}(\theta)$, given in Eq. (2.3), for the transitions $\psi \to \psi_{up}$ and $\psi \to \psi_{down}$. If we said "apparently" it is because, as we are going to explain, the extended Bloch model, if fully exploited, already contains all the information we need to deduce the above probabilistic formulas.

What is important to note is that we *do not know* how the breaking of the elastic, representative of the measurement process, exactly happens. We know that it breaks initially at one point, and that as a result of this breaking its collapse draws the corpuscle representative of the spin state into one of its two anchor points. But we do not know in which way the breaking is produced. In other words: *we do not know anything about the specific structure of the elastic and of the fluctuations that are causing its breaking.*

In fact, not only does the quantum theory, in its standard formulation, not give us any indication in that respect, but neither can we obtain any additional information by observing how an experiment like the one of Stern–Gerlach unfolds. Actually, the latter statement is only partially correct. Indeed, as the saying goes, *silence can be worth a thousand words*, in the sense that even a total absence of indications can provide a good indication, that is, a valuable piece of information.

What information is it? Well, the information that *an experimenter, when measuring a physical quantity, usually operates in such a way as to avoid as much as possible influencing the result of the measurement.* Since this is an important point, and at the same time a subtle one, let us try to explain what we mean exactly.

As mentioned above, a measurement is a process of the *interrogative* kind. To ask a question, however, is not an action without consequences, as is well known to psychologists. The answer one gets usually depends on how the question is asked, as well as on its context. For example, if we ask somebody if they like video games, the answer can vary considerably depending on who the person asking the question is. The answer is more likely to be negative when the question is asked by an employer of an employee than if it is asked between two trusted friends.

Of course, if we really want to know whether or not somebody likes video games, we have all interest in creating an interrogative context which will enable them to express themselves as *authentically* as possible, saying things exactly as they are, i.e., telling the *truth.* The same is true in physics when an experimenter measures a certain property of a physical entity. The experimenter's interest is to know if this property is truly possessed by the entity. This means that the question will be asked and the experiment will test the corresponding properties in such a way as to avoid any undue influence on the entity, since this could alter its outcome.

On the other hand, no matter how many precautions are taken in this respect, there is a minimum level of influence that experimenters will never be able to avoid. In fact, they need to ask a concrete question to obtain an answer, and to ask such a concrete question implies creating an experimental context that inevitably produces a change. What change? Well, simply, the change called "response to the posed question."

To ask a question is an action that stimulates the entity subjected to it to react in such a way that a specific answer to the posed question can be known, provided of course that the question is *well posed*. Certainly, in some circumstances it may happen that a question does not receive any answer, but anomalous events of this kind are not usually accounted for in the statistics of the results of an experiment. In the case of the Stern–Gerlach experiment, they may for example correspond to situations in which the silver atom, for whatever reason, fails to reach the detector (perhaps because it is absorbed by the magnet).

But let us explore a bit further the previous psychological analogy, in which a question is asked to a human subject. It is obvious that the stimulus of the question (the cognitive context it implies) will produce a change in the mental state of the subject, which will put it in the condition of being able to produce an answer. This change will be without major consequences if for the subject it is already clear (even before the question is addressed) what *the* answer is (and in that sense, the answer would be predictable with certainty by those who know the subject well).

Going back to the question about video games, a good experimental psychologist, really wanting to know the true answer to the question, will make sure that the person who poses the question, the way it is posed, the place where this happens and the very wording used, will allow the subject to respond with *maximum sincerity.* In other words, the investigator will try as much as possible *not to control* the response, allowing it to stem in a natural and spontaneous way from the subject, with no influence whatsoever.

On the other hand, not always will a person know in advance the answer to a question, even when it is about their own likings. While regular players of video games are very much aware of their appreciation of the games, there are undoubtedly many people who know with equal certainty that they do not like this form of entertainment. But there are also many people for whom the answer to the question is not at all obvious. Maybe because they have never played a video game, or have done it just a few times in their lives, for a short period, without being able to tell whether or not they enjoyed it.

Let us imagine that the question is asked to somebody from this group, i.e., somebody who has not yet formed an opinion, and that the only possible answers are "yes" or "no." Despite all the precautions that the interrogator may take to put the interrogatee at ease and encourage as much as possible an authentic response, the latter will literally have to *create* an answer of their own.

However, the result of this creation process will not depend in any way on the action of the interrogator. This is because, as we have assumed, every possible precaution would have been taken not to influence the outcome, creating a reassuring environment and avoiding asking the question in a biased way. For example, by avoiding formulations such as: "You are very smart, and I bet that you like video games, correct?" Or: "Video games are definitely an activity for time wasters, but tell me: do you like them?" More obvious formulations would aim to be as neutral as possible, such as: "Do you like video games?"

Therefore, ideally, the *indeterministic* (not predeterminable) *way* in which the person selects (creates) an answer, will not be dependent on the *deterministic way* in which the interrogator poses the question, because any form of control over the answer will have been thoroughly avoided in advance.

In the case of the quantum interrogations (measurements), the situation is exactly the same. The experimenter will create an experimental context that allows posing the question in a practical way. And being interested only in observing the answer, without influencing it, they will avoid bringing about any distortions in the experimental protocol. However, this will not be enough to eliminate any source of indeterminism in the answer. In fact, the measured entity will not necessarily be in such a state that the value of the observed quantity is already actual. And if it is not, as is usually the case, when subjected to the action of the experimental context (i.e., when "obliged" to answer), the quantity in question will have to be created (actualized), and this process, for its very nature, will be subjected to fluctuations that make it inherently unpredictable.

Let us be clear. The investigator is certainly free to design an experiment such that its outcome is certain in advance, regardless of the initial state of the entity subjected to it. For example, in the case of spins, the silver atoms may be submitted to the action of specific "electromagnetic lenses" able to align them all along a predetermined direction, irrespective of their initial direction (Karimi, 2012). This type of process, however, cannot be considered a measurement of the spin orientation. Rather, it would correspond to a situation in which the experimental psychologist from the example given above asks the question "Do you like video games?", while holding a gun and adding "if you say 'no' I'll shoot you."

This kind of experiments only show that spins and humans share the special property of being always "orientable," when subjected to the appropriate experiments, but they are certainly not experiments aiming to observe the spatial orientation of a spin, or of the "ludic orientation" of a person.

After this long digression, we can now return to our *extended Bloch model*. The Stern–Gerlach experiment is a measurement of the orientation of the electron's spin and, as we explained, as a measurement it corresponds to a situation where the experimenter has eliminated all forms of possible control over the orientation of the electron's spin. The experimental *set-up* corresponds to a question posed in a way that is as neutral and non-invasive as possible; the Stern–Gerlach apparatus will not know in advance the answer to the question, when the spins are prepared along other directions than an initial "up" or "down" orientation, so that it will have to select an orientation, that is, to create the "up" or "down" final state, corresponding to the answer.

How does this process of creation come about? How does the electron spin collapse its initial state into an "up" state or a "down" state? Well, as we said, *we know nothing about this*! But nevertheless, we know something important: that the experimenter will do anything to not affect the way in which the final state is selected.

The measurement process is a process of observation precisely because it doesn't force the apparatus to create the final state in one way rather than another. It is free to do this "as it thinks best," only on the basis of the random fluctuations that are present in nature, that is, that are *present when no attempts whatever are made to control such fluctuations*.

In the extended Bloch model, these fluctuations are represented by the structure of the elastic band: *every measurement process corresponds to the choice of a specific elastic*. However, the elastic is not chosen by the experimenter (otherwise such information would be available): it "emerges" in a totally natural way, that is, spontaneously and unconditionally, as a consequence of the environmental fluctuations. This means that there are as many *ways to answer* the question asked by the experimenter — "What is the value of the spin along the direction of the magnetic field?" — as there are possible breakable elastics that, by breaking, are able to produce an answer.

The attentive reader may have begun to connect what we are explaining here with our previous analysis of Bertrand's paradox. As we pointed out, the hard part of the paradox, once its question has been disambiguated, is to calculate an average over all the possible ways to throw a straw onto a circle, to get a chord the length of which could be longer or shorter compared to the side of the inscribed equilateral triangle.

So, the question posed in Bertrand's problem can also be interpreted as the measurement process of a *non-ordinary* property of the straws — let us call it *equilaterality* — by throwing them onto a circle drawn on the floor, a throwing that is part of the observational protocol of such property.

The observation of the equilaterality of the straw can only give two outcomes: "longer" or "shorter," in the same way that the observation of the electron's spin (along the magnetic field direction) can only give two outcomes: "up" and "down." Now, the very definition of the equilaterality of the straws, as discussed previously, does not allow for any control by the experimenter. In fact, in his question, Bertrand did not specify *how* the straw had to be thrown, but only that the straw *had* to be thrown.

Similarly, when we measure the orientation of the electron's spin, the protocol does not specify *in which way* the apparatus has to actualize the final spin orientation, that is, in which way the elastic band within the Bloch sphere has to break, but only that it will have to break (i.e., produce an answer). In other words, *we do not know and do not control* the nature of the elastic that each time produces an outcome.

Therefore, in the same way that to seriously answer Bertrand's question we must consider all possible ways of throwing a straw, to derive the experimental probabilities in the Stern–Gerlach experiment we have to consider all the possible ways that an elastic can break. This is because, seeking neutrality, the experimenter has done everything possible to remain *meta-indifferent* with respect to all these ways, since favoring some of them over others would mean a conditioning of the outcome of the experiment, which in this case would no longer be interpretable as an observational experiment, that is, as a measurement.

So, similarly to what we did in the Bertrand's problem situation, we must consider all possible types of breakable elastics, calculate for each of them the probabilities of getting ψ_{up} and ψ_{down}, then work out the arithmetic mean of all these probabilities, and finally compare it with the value predicted by the Born rule. If the value is the same, this means that we have been able to derive the Born rule by considering it as an expression of a *universal average*. Consequently, the quantum measurements would be nothing more than *universal measurements*!

Before considering how we can work out this calculation, it is instructive to analyze certain types of breakable elastic, among the *infinite number* of possible ones. The simplest case is that of an elastic that can break at *only one predetermined point*. Elastics of this kind model *deterministic* processes. In fact, knowing the initial state ψ, we can predict in this case with certainty whether the final state (i.e., the outcome of the measurement) will be ψ_{up} or ψ_{down}. This is because, as stated above, the preliminary process of immersion of the particle inside the sphere is perfectly deterministic (it is a movement of "orthogonal fall" onto the elastic). Therefore, the exact "landing" point only depends on the value of the angle between the orientation of the initial state ψ and the orientation of the elastic band (see Figure 3.5).

More precisely, if we put the points of the elastic into a correspondence with the interval of real numbers $[-1, 1]$, where -1 corresponds to the anchor point ψ_{down}, 1 to the anchor point ψ_{up}, and 0 to the

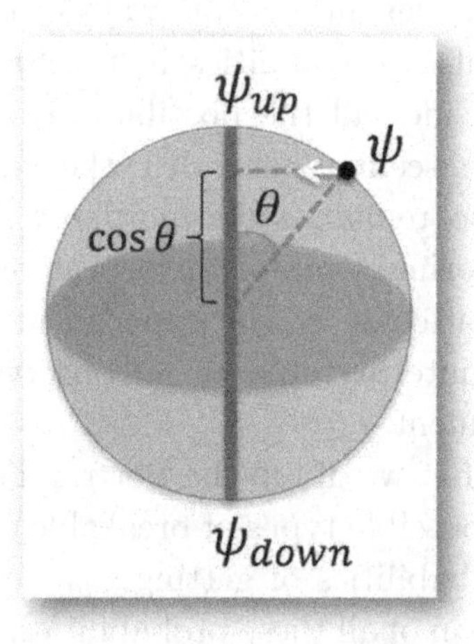

Fig. 3.5 The corpuscle, when it "falls" onto the elastic, lands at a point at a distance $\cos\theta$ from its center.

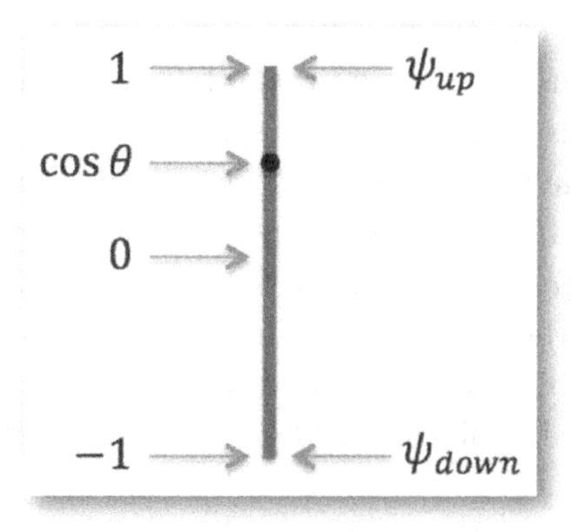

Fig. 3.6 The corpuscle "lands" on the elastic at the point $\cos\theta \in [-1, 1]$.

center of the elastic (which is also the center of the sphere), then, being the sphere of radius 1, the corpuscle, once "stuck" to the elastic, will be exactly at the position corresponding to the value $\cos\theta$ (see Figure 3.6).

So, if we know that the elastic is of the type "*that breaks only at point* x," then we also know that if the initial state is such that x is strictly less than $\cos\theta$ ($x < \cos\theta$), with certainty the corpuscle will end its run at the position ψ_{up}. This is because it will be located on the elastic between ψ_{up} and the breaking point, that is, on the fragment that will collapse towards ψ_{up}.

Similarly, if the initial state is such that x is strictly greater than $\cos\theta$ ($x > \cos\theta$), with certainty the corpuscle will end its run at the ψ_{down} position, being located on the elastic between ψ_{down} and the breaking point, i.e., on the fragment that will collapse towards ψ_{down}.

What happens, though, if the initial state is such that $x = \cos\theta$, namely if the corpuscle lands exactly at the breaking point? In this eventuality, it is of course no longer possible to predict the outcome of the measurement, as the corpuscle will have equal probability of being drawn by either fragment, or else it could simply remain at the center of the sphere without producing any specific outcome.

If we take into account these quite exceptional events (that in any case do not contribute to the final statistics, since they are, as mathematicians call them, *events of zero measure*), we should say that the elastic bands that only break at one point describe processes that are *almost* deterministic, in the sense that they are deterministic for *almost* all of the initial states, except those exceptional situations where the particle falls exactly on the predetermined breaking point, giving rise to a purely indeterministic process.

Consider now the situation in which the elastic can break only at two points, which we will call x and y, with y greater or equal to x ($-1 \leq x \leq y \leq 1$). In this case, the dynamics is more complex, since it is possible to distinguish three situations. The first corresponds to the particle "landing" on the elastic at a point between -1 and x ($-1 \leq \cos\theta \leq x$). The process is then deterministic, and with certainty we know that the outcome will be ψ_{down}, as both breaking points are located between the particle and the anchor point ψ_{up}. Similarly, if the "landing" point is between y and 1, the outcome will certainly be ψ_{up}.

The situation is different when the contact point is located on the elastic segment between x and y (see Figure 3.7). In this case, it is not possible anymore to predict with certainty the outcome of the measurement, which depends on which of the two points will break. If the probability for point x to break is P_x, and the probability for

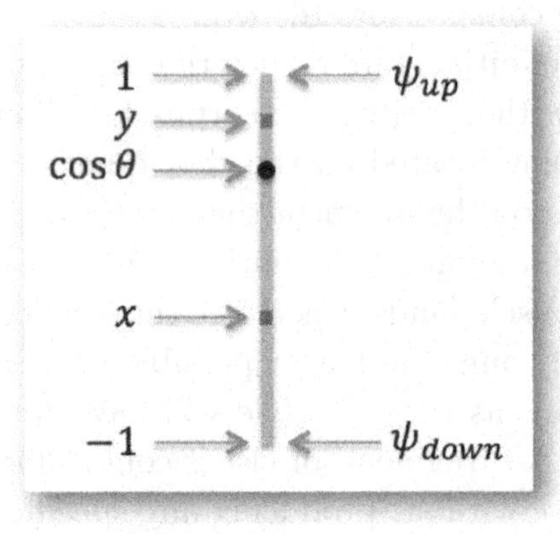

Fig. 3.7 An elastic that can only break at two points, x and y, gives rise to an indeterministic process, if the initial state is such that $x \leq \cos\theta \leq y$.

point y to break is P_y, then the probability for the outcome ψ_{up} is P_x, and the probability for the outcome ψ_{down} is P_y.

What happens if $x = -1$ and $y = 1$? In this case, the probabilities for the outcomes ψ_{up} and ψ_{down} no longer depend on the angle θ. In other words, whatever the initial state of the measured entity is (the initial position of the corpuscle on the surface of the sphere), the outcome of the measurement will not depend in any way on it (apart, of course, from the exceptional situations $\cos\theta = 1$ and $\cos\theta = -1$, where the spin is already in an eigenstate). This is an extreme case, characterizing a measurement that is called *solipsistic*.

Solipsism is the philosophical view holding that everything an individual observes is autonomously created by his own consciousness, rather than depending on the intrinsic properties of the observed entity. By analogy, a solipsistic measurement is a measurement whose result does not reveal anything about the initial state of the measured entity, as the probabilities are only determined by the particular structure of the elastic. Therefore, we can say that opposite to the deterministic processes, which maximize the *discovery* aspect, the solipsistic ones maximize the *creation* aspect.

Returning again to the example of the video game question, we can say that a solipsistic measurement would correspond to a context in which the interrogatee is not at all sensitive to the way the question is asked, not because the answer is predetermined, but because it is impossible to understand all the nuances of the language used to formulate it, the different symbolic and figurative contents, etc. As a result, instead of revealing something about the nature of the question, and the way it was formulated, the answer only reveals the nature of the interrogatee's mental state at that specific moment.

Another example of solipsistic measurements are those used in the so-called *games of chance*, such as *craps*, played mainly in America and Canada, where one bets on the outcome of the roll of two dice. The protocol requires the player to toss the dice with vigor, using only one hand, making them bounce on the far wall of the table. In this way, the initial position of the dice in the thrower's hand (their initial state) is in no way permitted to influence the outcome of the roll (their final state), i.e., the value of the observable "upper face of the dice." In other words, the probability of the different outcomes does not depend in any way on the initial condition of the dice, and this solipsistic property of the throwing protocol is precisely what allows the game to remain fair.

That said, returning to our representation in the Bloch sphere, if we take into account every possible type of breakable elastic, in what we have defined to be a *universal measurement*, we will have to deal with every possible experimental context, that is, with every possible context in terms of discovery and creation, compared to borderline cases of the classical contexts, where the discovery aspect is maximized, and of the solipsistic contexts, where it is the creation aspect that is maximized. So, if we can show that the universal measurements are equivalent to the quantum measurements, these can be understood as measurements that realize a kind of *balance* between the two fundamental and complementary aspects of discovery and creation, which are always present in an observation-measurement process.

Now, to describe every possible type of breakable elastic, it is sufficient to associate every elastic with a different probability distribution ρ, able to attribute to each of its infinitesimal segments a specific probability of breaking in the course of an experiment. The probabilities associated with a universal measurement then correspond to the uniform average over the probabilities associated with all possible and imaginable elastic bands, characterized by all possible probability distributions ρ.

Let us try to be a bit more specific. To make a universal measurement means to do the following:

(1) For every measurement, we must first *prepare* the entity to be measured in a given state, which corresponds to a specific point on the Bloch sphere;
(2) We must then *randomly select* an elastic band among all possible elastic bands, i.e., a probability distribution ρ;
(3) We must next use the chosen elastic, characterized by the probability distribution ρ, to conduct the experiment by means of the indeterministic breaking process and the subsequent collapse;
(4) Finally, we have to record the obtained result.

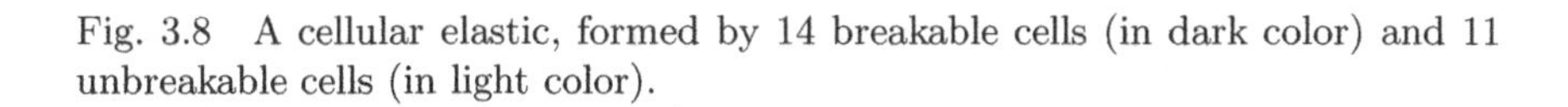

Fig. 3.8 A cellular elastic, formed by 14 breakable cells (in dark color) and 11 unbreakable cells (in light color).

To find the probabilities, we must obviously repeat the four above-mentioned operations many times, then calculate the relative frequencies of the different outcomes, which will correspond to the probabilities characteristic of a universal measurement, which we can compare with the quantum probabilities defined by the Born rule.

Here we find ourselves in the same situation that we have described in relation to Bertrand's paradox. The probability distribution ρ, rather than characterizing the different *ways* to throw a straw on a circle, characterizes the different *ways* that an elastic can break when used to perform a measurement in the Bloch sphere.

Thus, similarly to what we have previously explained, in order to take the average over all the possible elastics, breaking in all possible ways, the winning strategy is to consider *cellular elastics* characterized by only two types of cells: those that are breakable (all with the same probability) and those that are unbreakable (see Figure 3.8).

It is then possible to show, as in the analysis of the hard part of Bertrand's problem, that each probability distribution ρ, associated with an arbitrary elastic, can be approximated with arbitrary precision by a sequence of cellular elastics, to the extent that the number n of their cells increases. Therefore, we can first take the average over the probabilities of all possible elastics formed by exact cells, and then take the limit $n \to \infty$ (n tends to infinity) to finally obtain the probabilities associated with a universal measurement.

Similarly to what we have obtained in the solution of Bertrand's paradox, if we do this two-step calculation (the arithmetic mean over all cellular probabilities and the limit $n \to \infty$), the result that we obtain is that a universal measurement is equivalent to a measurement made by using a single *uniform elastic*, that is, an elastic having the same probability of breaking at all of its points (characterized by a *uniform* probability distribution ρ_u).

It is certainly an expected result (although not for the obvious reason), for reasons of symmetry, as it is easy to imagine that when we do the average over all the possible elastic bands, the different inhomogeneities of the individual elastics can compensate one another, so as to produce an *effective homogeneous elastic*, that is, a uniform one, which well expresses our condition of *meta-ignorance* (ignorance not only about the nature of the elastic, but also about its breaking point).

At this point, we only have to verify that a measurement process that uses a uniform elastic band is able to reproduce the same probabilities predicted by the Born rule. The calculation is easily done. The elastic being uniform, the probability that when it breaks the corpuscle representative of the state is dragged towards ψ_{up}, corresponding to the position $x = 1$, is simply given by the ratio between the length L_{up} of the elastic fragment between the position of the corpuscle, at $x = \cos\theta$, and the opposite anchor point, at $x = -1$, and the total length L of the elastic, as the points forming L_{up} are precisely those that, when they break, produce the outcome ψ_{up} (see Figure 3.9).

So, considering that the total length of the elastic is $L = 2$ (twice the unit radius of the Bloch sphere), we obtain:

$$P_{up} = \frac{L_{up}}{L} = \frac{\cos\theta - (-1)}{2} = \frac{1 + \cos\theta}{2}, \tag{3.1}$$

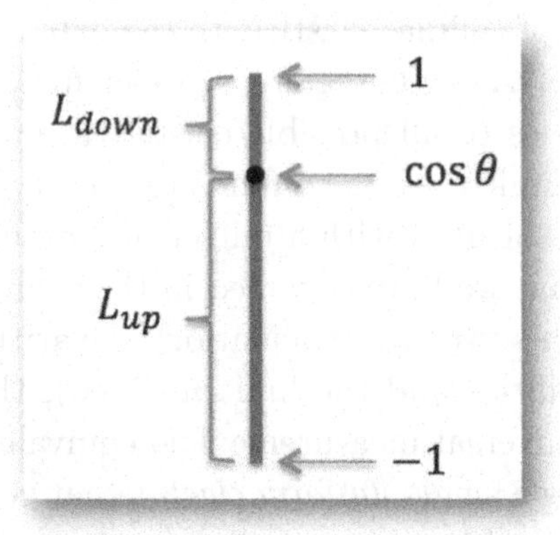

Fig. 3.9 The point particle representative of the state, once "landed" on the elastic, defines two segments of length L_{up} and L_{down}, respectively.

which is exactly the value predicted by the Born rule (2.3). And of course, *mutatis mutandis*, the probability for the outcome ψ_{down} is given by:

$$P_{down} = \frac{L_{down}}{L} = \frac{1 - \cos\theta}{2}, \tag{3.2}$$

which again is in perfect agreement with the Born rule (2.3).

We have thus succeeded in deriving the Born rule starting from the Bloch sphere representation of the quantum states, and by modeling the measurement process by means of a (non-ordinary) disintegrable, elastic and attractive substance applied along the two antipodal points of the sphere, corresponding to the two possible outcomes of the measurement. We have done this by only exploiting the available experimental information, that is, without assuming any *a priori* knowledge about the dynamics of disintegration of the elastic, in agreement with the fact that the experimenter does not impose any type of *a priori* control over the outcome of the experiment.

In other words, by exploiting the notion of *universal average* and the Bloch representation, we have offered a possible *solution to the measurement problem*, showing (here in the case of spin entities) that the Born rule is compatible with the predictions of a universal measurement. In other words, the Born rule can be understood as a universal measurement acting on the specific structure of the state space of quantum mechanics (the Hilbert space), where the Bloch representation is completed with the internal points of the sphere.

Chapter 4

Measurements with N Outcomes

What we explained in the previous chapter is of course valid for a physical entity whose measurements can give only two possible outcomes, as is the case for the spin of an electron, called spin $\frac{1}{2}$ (spin one-half), since its observation, in a given direction, can produce only two values: $+\frac{\hbar}{2}$ ("up" state) and $-\frac{\hbar}{2}$ ("down" state), where $\hbar$ (pronounced "h-bar") is the famous *Planck constant*, in its reduced version (also called *Dirac constant*).

In the psychological analogy, the measurement of a spin $\frac{1}{2}$ is the equivalent of an interrogative context that admits only two answers, for example "yes" and "no," usually associated in physics with the so-called *two-level systems*, also named *qubits* (quantum bits) or just *two-dimensional systems*, since the corresponding Hilbert space has exactly two complex dimensions (whereas the associated Bloch sphere has three real dimensions).

At this point, it is natural to wonder if what we have so far explained remains valid also for more general observational contexts having an arbitrary number N of possible outcomes. In fact, one might suspect that two-dimensional systems, which only allow two outcomes, are something of an anomaly, where a "hidden-measurement explanation" would indeed be possible, but that this explanation would not be working anymore if the number of outcomes exceeds two.

To understand the reasons for this possible objection, it is worth remembering that during the development of quantum physics there have been numerous attempts to complete this admirable theory, by adding variables in order to obtain a more accurate description of the state of a physical system and thus explain quantum indeterminism as a *condition of lack of knowledge about the state of the system.*

Indeed, since the results of experimental observations could only be predicted in probabilistic terms (apart from exceptional cases where the initial state is an *eigenstate*), this could mean that we were missing some important information about the properties of a quantum entity. In other words, the assumption was that there were *hidden variables*, yet to be identified, the determination of which would allow one to build a more complete theory, this time perfectly deterministic.

These attempts to construct more complete quantum theories are today known by the name of *hidden-variables theories.* Over time, however, they have clashed with some insurmountable obstacles: the so-called *no-go theorems*, or *impossibility theorems*, according to which the "classic dream" of coming back to purely deterministic measurement processes was gone forever.

In fact, these impossibility theorems unequivocally showed exactly that if you added variables to the states of a quantum entity, irremediably the probabilities associated with these new "completed states" would be of the classical kind, and not of the quantum kind. But this was a problem because, as we mentioned in the course of our analysis of Bertrand's paradox, the quantum probabilities, associated with the Born rule, violate the axioms of classical probability theory, as enunciated by Kolmogorov. So, the assumption underlying the traditional hidden-variables theories could not be correct.

Let us take advantage of this brief digression to better emphasize the fundamental difference that exists between our *extended Bloch representation*, which is based on the idea of the *hidden measurements*, and the "too classic" idea at the basis of the historical hidden-variables theories, according to which, instead, the "hidden" element should only be associated with the states of the measured physical entity. To this end, it is useful to recall what *Claude Bernard* (the father of scientific physiology; see Figure 4.1) defined as the *absolute*

Fig. 4.1 C. Bernard.

principle of the experimental method. This principle simply affirms that (Bernard, 1949):

If an experiment, when repeated many times, gives different results, then the associated experimental conditions must have been different each time.

Clearly, this is a reversed, alternative way to state the famous *principle of determinism*, affirming that if everything is given in an experiment, then there are no known reasons to think that the result of the experimental process, if properly conducted, wouldn't be predetermined, whatever the outcome will be. But what is particularly interesting in the enunciation of the principle of the experimental method, compared to the principle of determinism, is that Bernard emphasized the fact that the experimental context can change, producing different outcomes. In other words, its principle leads us to reflect on the nature and origin of the *fluctuations* that can be present in an experimental context.

These fluctuations could certainly be about the state of the measured entity, according to the assumption underlying the traditional hidden-variables theories, which then collided with the aforementioned impossibility theorems. On the other hand, the possibility also exists that the fluctuations are to be attributed, rather than to the state of the entity, to the dynamics of its interaction with the measuring system. And this is exactly the idea behind the *hidden-measurements interpretation of quantum mechanics*, which is incorporated in the extended Bloch model.

Among the reasons why this idea was not initially taken into sufficient consideration, there is the *classic prejudice* that an act of observation should always be predictable, when we have complete knowledge of the state of the observed system, and that such an act, by definition, should always be considered to be of a *non-invasive* nature, or that in any case it could always be made such with the appropriate technical solutions. This bias, however, is not only contradicted by the observations experimental physicists execute in their laboratories on a daily basis, when observing the properties of microscopic entities, but also by our most ordinary observations, when we reflect more carefully about what it means *to observe a property.*

For example, let us imagine that we want to observe the *burnability* of a wooden cube. Of course, to do so, we must execute an *observational test* that consists in exposing the wooden cube to a flame, so as to check if it can really burn (see Figure 4.2); and certainly, if it proves to be burnable, the observation will prove to be quite invasive, as it will cause the partial or total destruction of the observed entity, something that is inevitable if we want to measure this property directly.

In other words, the invasiveness of certain observational processes is inherent in the very nature of the observed properties, that is, in the observational protocols through which the properties are defined in *operational* terms, relative to the entity in question. The last point is quite important, as the observation of the same property can prove to be invasive for a given entity, and non-invasive for an entity of a different nature.

Consider the example of measuring the orientation of a magnetic dipole. When it comes to a macroscopic magnet, we can always perform the measurement in a non-invasive way; whereas if the dipole

Fig. 4.2 Observation of the burnability of a wooden cube.

in question is associated with the spin of a microscopic entity, like an electron, it becomes impossible, in general, to not change the state of the system through the measurement.

Since this is a delicate point, and at the same time a crucial one, we open a brief parenthesis to ask why this happens. If we go back to our analysis of the Stern–Gerlach experiment, we can observe that when the deflecting magnet is oriented along the same direction as the initial spin of the electron (to which is associated the corresponding magnetic dipole), the measurement will confirm with certainty such orientation. As we have already explained, an initial state of this kind (with the same spatial orientation as the measuring magnet) is called an *eigenstate*, and we can say that in this type of situations the measurement is perfectly non-invasive, as it does not produce a change of the state (i.e., of the orientation) of the observed entity.[1] This means that, knowing in advance the eigenstate of the electron, we can in principle always make sure that the measuring instrument is oriented along the same spatial direction, and thus produce a non-invasive observation, the result of which is *a priori* certain.

In other words, it could be argued that the invasiveness of the measurement results from the fact that we measure the wrong observable, because it is not oriented along the direction of the eigenstate, and that it is precisely this mismatch which produces the unpredictable change of the state. This possible objection doesn't tell us, however, what would be the mechanism that makes the process invasive, as well as unpredictable. Also, it is based on a generally invalid assumption: that it is always possible to associate a spatial direction to each spin state.

We have already observed that an electron has a spin $\frac{1}{2}$, and the same goes for a neutron, a proton, and a bunch of other elementary entities. As we have seen, in the case of a spin $\frac{1}{2}$, the state space is a three-dimensional (Bloch) sphere, as our *Euclidean spatial theater* is three-dimensional. And since to every state of the electronic spin we can associate a point on the sphere, and to every point on the sphere we can associate a spatial direction, it follows that to each electronic spin state we can associate a direction of the three-dimensional physical space: exactly the direction for which the spin state is an eigenstate.

In short, when we are dealing with spin $\frac{1}{2}$ elementary entities, each spin state is also, necessarily, an eigenstate for a given direction

---

[1]If we consider the Stern–Gerlach experiment in its entirety, this statement is not totally correct, as the silver atom is ultimately absorbed by the screen end-detector, and this absorption process is undoubtedly of an invasive nature. However, what interests us in our reasoning is the change that the measurement process induces in the spin state, regardless of the ultimate fate of the atoms subjected to the process.

of space. But is it always so for every spinorial entity? The answer, surprisingly, is negative, and in this sense spins with value $\frac{1}{2}$ are to be considered as genuine anomalies.

We recall that in nature there are entities with different spin values. In fact, although it is true that all elementary *fermions* (which form the so-called "ordinary matter") have a spin $\frac{1}{2}$, when they combine with each other they can create composite entities, whose spin can take any (integer or fractional) value among the permitted ones: $1, \frac{3}{2}, 2, \frac{5}{2}$, etc. Also, the so-called elementary *bosons* — associated with force fields, like the *photon* (electromagnetic force), the *gluon* (strong force) and the W and Z *bosons* (weak force) — have a spin 1, and many scholars believe that the hypothetical *graviton* (gravitational force) possesses a spin 2.[2]

Now, in the case of an entity of spin 1 or higher, the infinite majority of states in which it can be are not eigenstates. In the sense that it is no longer possible to associate to each state a specific direction in space, such that a Stern–Gerlach magnet, if oriented along that direction, would produce a result certain in advance. In other words, excluding the anomaly of the spin $\frac{1}{2}$ entities, the eigenstates of a generic spin observable only represent an infinitesimal fraction of the possible states. And this means that in general, before a measurement process, we cannot associate any spatial orientation with a spin. Thus: *it is the very measurement process that creates the spatial orientation*!

If this interpretation is correct, the measurement process can in no way be understood as the *disturbance* of an entity whose spin would initially be oriented along a predetermined direction and that because of the interaction with the magnet it would undergo a modification, but as the *creation* (in the literal sense) of a *spatial direction for the spin* — a direction that was absent (not existing) before the observational process. And of course, a process of creation, by its very nature, can only be invasive, and certainly cannot be made less invasive by using whatever technical expedient.

[2]Apparently there are also *scalar* elementary entities with spin 0, as is the case of the famous *Higgs boson*, which recently has probably been detected.

From a mathematical point of view, the existence of these *non-spatial states* manifests very clearly in the fact that the Bloch sphere is not three-dimensional any more, but of a higher dimensionality. For example, in the case of an entity of spin 1, whose measurements can give rise to three distinct outcomes, the representation of all possible states requires a *hypersphere* having eight *dimensions.*[3] It should also be said that the states of a spin $\frac{1}{2}$ entity, although representable in a three-dimensional Bloch sphere, if carefully analyzed, also reveal a non-spatial nature. In fact, we cannot deduce the totality of their properties, in the different experimental contexts, from the simple hypothesis that a spin would be describable as a classic three-dimensional real vector.[4]

But in the case of a spin 1 (or of greater value) entity, the non-spatiality of its states is totally evident. In fact, it is possible in this case to identify within the eight-dimensional sphere those vectors that represent the different spatial directions (which form a three-dimensional subsphere), and show that none of the spin states is actually oriented along a spatial direction, not even the eigenstates (Aerts & Sassoli de Bianchi, 2015a).

In the case of spin 1, the number of eigenstates relative to a given direction of measurement is three. In other words, when we measure a spin 1 entity with a Stern–Gerlach apparatus, we can get three distinct outcomes: a state ψ_{up}, associated with the maximum value of the spin, which is $+\hbar$; a state ψ_{down}, associated with the minimum value of the spin, which is $-\hbar$; and a state ψ_0, associated with a *null value* of the spin (corresponding to an absence of deviation by the magnet). These three possible outcomes are associated with three different probabilities, P_{up}, P_{down} and P_0, which depend on the state of the spin entity prior to the observation, and these probabilities

[3]Actually, only a *convex subset* of this eight-dimensional sphere contains the states of the entity, but this is a technical detail of which we will not have to worry here.
[4]It is known for example that when a spin $\frac{1}{2}$ is rotated by 360°, it will not be back exactly in the same state, as for this one needs to execute a 720° rotation. Obviously, such a property is hardly attributable to an entity entirely representable in space (see for example Sassoli de Bianchi (2016) and the references therein).

can of course be calculated using Born's "golden rule." However, the formulas being much more complicated in this case, we will not write them explicitly.

What interests us is to understand how to describe this experimental situation with three possible outcomes, generalizing the previously described process for a spin $\frac{1}{2}$ (with only two outcomes). Of course, in this case we will not be able to draw the whole state space, since it is eight-dimensional; nevertheless, we will be able to visualize the *potentiality region* that characterizes the interaction between the measured entity and the measuring instrument. In the case of spin $\frac{1}{2}$, this region corresponded to the action of an attractive, elastic and breakable *one-dimensional* structure, stretched between the two points representative of the two possible outcome states.

In this case, however, having to do with three possible outcomes, the structure in question can no longer be one-dimensional, but has to be *two-dimensional*. In other words, it is in this case a *membrane*, the shape of which is precisely that of an *equilateral triangle* inscribed in the eight-dimensional Bloch sphere, the vertices of which correspond, respectively, to the three possible final states: ψ_{up}, ψ_{down} and ψ_0.

In this case, we can also distinguish two stages in the measurement process.[5] The first always corresponds to a purely deterministic process, during which the material particle plunges into the Bloch sphere to reach the membrane in a rectilinear "falling" movement, i.e., *perpendicularly* to the plane of the same. This immersive movement stops as soon as the corpuscle touches the membrane, firmly attaching to it, at which point its attraction is maximal (see the first drawing of Figure 4.3).

Once the corpuscle is on the membrane, its presence determines three *triangular subregions*, delimited by particular "*tension lines*" along which the membrane has greater difficulty to disintegrate. More

[5]In the more general situation of measurements that are called degenerate, where the experimenter cannot distinguish some of the outcomes, one must add a third stage to the process. Not to complicate the presentation, we will omit the description of these types of measurements and refer the interested reader to the articles cited in the bibliography.

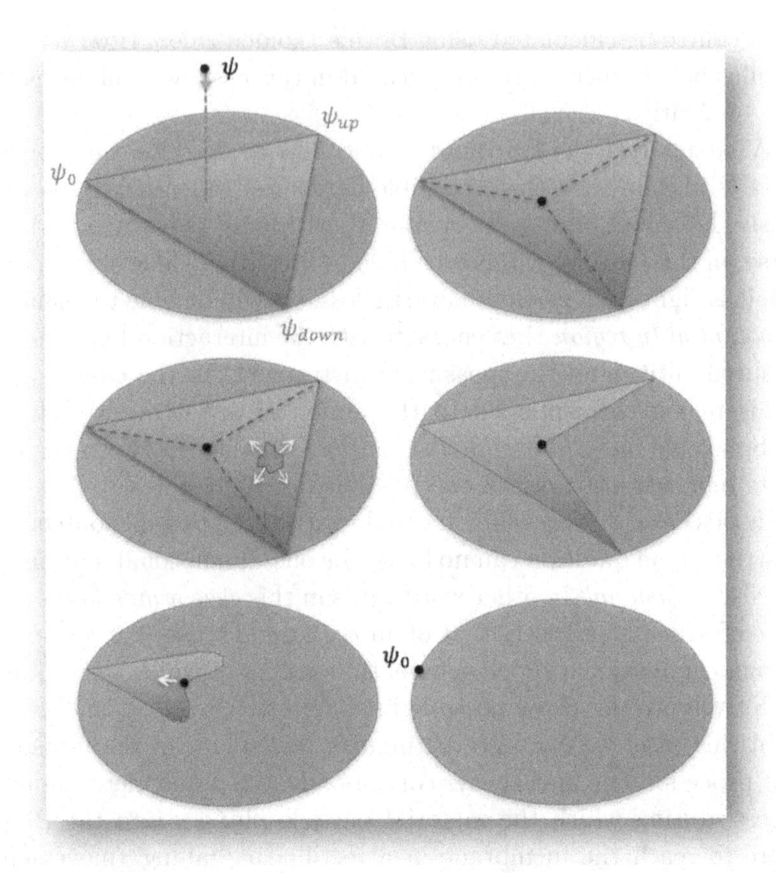

Fig. 4.3 The measurement process of a spin 1 entity. The material point representative of the initial state ψ is initially located at the surface of the eight-dimensional hypersphere, not shown in the drawing. From this position, it plunges inside the hypersphere along a trajectory that is perpendicular to the plane of the membrane representative of the measuring system. The subsequent breaking of the membrane, at a point belonging to one of its three subregions defined by the "tension lines" generated by the corpuscle, produces the disintegration of that subregion, and the contraction of the membrane towards the remaining anchor point, thus determining the outcome of the measurement, in this case ψ_0.

precisely, these are the lines connecting the position of the particle to the three vertices of the triangle (see the second drawing of Figure 4.3).

Now, because the membrane is *unstable*, sooner or later it will disintegrate, at some unspecified point belonging to one of these three subregions, starting the second stage of the measurement process, which is generally *indeterministic*. This corresponds to the propagation of the disintegrative process within the subregion where it originated, with the consequent loss of its two anchor points to the sphere; and because the membrane is *elastic*, its undocking will cause its contraction, that is, its collapse, towards the only remaining anchor point, bringing the corpuscle attached to it in this position, which then corresponds to the final state of the spin (see the last four drawings of Figure 4.3).

At this point, we can reason exactly as we did in the case of the two-outcome measurements. The probabilities of the different outcomes, in fact, will depend on both the position of the particle on the membrane (which in turn depends on its initial position on the generalized Bloch sphere), and on the characteristics of the membrane, described by a particular probability distribution ρ. And since we have no specific information about the characteristics of the membrane that is selected at each run of the experiment, we have to perform a *universal average* over all possible membranes, and it is always possible to demonstrate that this is once again equivalent to an *effective membrane of the uniform type*, having the same probability of breaking at each one of its points (described by a *uniform* probability distribution ρ_u), which expresses the experimenter's condition of meta-ignorance.

We can then determine the different probabilities by simply calculating the ratio between the area of the three subregions and the total area of the triangular membrane. This is a relatively simple calculation of *analytical geometry* that, if done carefully, allows demonstrating that the membrane's mechanism determines *exactly* the probabilities predicted by the quantum-mechanical Born rule. In other words, the mechanism allows to *derive* the Born rule and thus to explain its possible origin — the rule describes an experimental situation where fluctuations are present at two distinct levels, generating two distinct mechanisms of "symmetry breaking": (1) in selecting a specific membrane among all the possible ones, and (2) in selecting a specific breaking point among all the possible ones.

Before concluding this section, some observations are in order. First, it is important to specify that the process that we have described — with three possible outcomes — is easy to generalize, in the sense that the same type of mechanism also works for a measurement with N possible outcomes, where N can be an arbitrarily large number. The membranes then become *hypermembranes*, forming $(N-1)$-dimensional *regular simplexes*,[6] inscribed in generalized Bloch spheres having $N^2 - 1$ dimensions. In the $N = 4$ case, for example, the shape of the membrane is that of a *tetrahedron* (see Figure 4.4), whose four vertices correspond to the four possible outcomes. And of course, in the general case, it can also be shown that

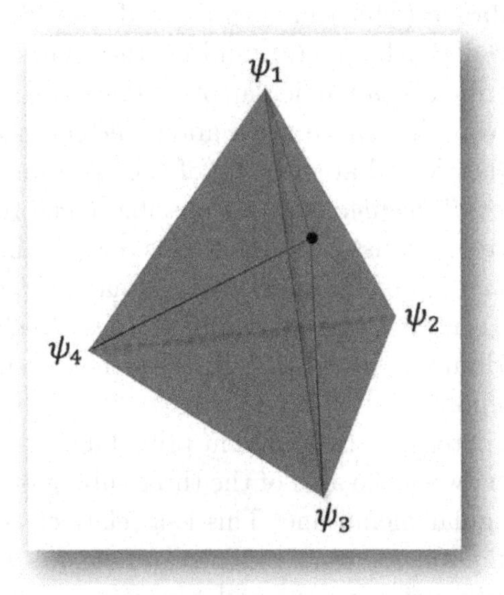

Fig. 4.4 In the case of a measurement with four possible outcomes, the potentiality region that realizes the transition from abstract probabilities to concrete actualities is a tetrahedron.

---

[6]A simplex is the generalization of the notion of a triangle. A one-dimensional simplex is a simple *line segment.* A two-dimensional regular simplex is an *equilateral triangle.* A three-dimensional simplex is a *tetrahedron.* A four-dimensional simplex is a *pentachoron.* And so on.

the relative *Lebesgue measures*[7] of the different subregions precisely correspond to the probabilities predicted by the Born rule.

Thus, the "hidden-measurements mechanism" described by the collapse of the membranes, being valid for an arbitrary number N of dimensions, i.e., of outcomes of a measurement, cannot be considered a mere two-dimensional anomaly, but the expression of a possible universal mechanism that can solve the measurement problem of quantum physics. Of course, in the present state of our knowledge, this remains a theoretical model awaiting experimental confirmation, since we have not yet been able to put in evidence the existence of the *measurement interactions* represented in our model by the membranes and their breaking mechanism. These are undoubtedly *non-ordinary* interactions, of a *non-spatial* kind, acting at the *sub-quantum* level, but this does not mean that well-thought experiments cannot reveal them in a near future. In short, the ball is now in the court of the experimentalists.

We have already evoked the concept of *non-spatiality*, on which there would be of course much to add, but this is not the subject of our booklet. The extended Bloch model, and the associated hidden-measurements interpretation, offer us a vision of reality where the so-called *quantum non-locality* can be reinterpreted as *quantum non-spatiality*, in the sense that, although our three-dimensional Euclidean space is certainly a theater for the macroscopic ordinary entities, it is not for this the container of all physical reality, since microscopic quantum entities, when they are in non-local states called *superposition states* (which are the majority), cannot anymore be described as belonging to this theater.

When it comes to space, by association *time* immediately comes to mind. What about the temporal aspects of the quantum entities? If these are generally outside of the three-dimensional Euclidean space, can we say that they are also outside of the four-dimensional

[7]The *Lebesgue measure* — named after the French mathematician *Henri Lebesgue* — is the generalization of the notions of *length*, *area* and *volume* to an arbitrary number N of dimensions, and is sometimes simply called N-volume.

spacetime? It is certainly possible: as well as a measurement process, conducted in a given *spatial location*, is able to spatialize a non-spatial quantum entity, we cannot exclude that the same process, unfolding in a given *time interval*, is not additionally able to temporalize an entity which is generally also *non-temporal*.

In this regard, it is important to observe that, even though our model makes use of a time-like description — with the material point that first plunges into the sphere, then reaches the membrane, and is subsequently dragged by its collapse towards one of the vertices — we do not know if such dynamics really takes place in time, for instance in a very swift way, or if the entire description is only a convenient way to represent a genuinely *atemporal* "process," which from our perspective would take place in a completely instantaneous way.

In other words, our model is not inconsistent with the hypothesis that quantum entities are not only non-spatial entities, but also non-temporal entities, as the evolution of the abstract corpuscle within the Bloch sphere does not necessarily happen in time (at least not in time as commonly understood). It could very well be a non-spatiotemporal process that we humans, for convenience and because of our limitations, would "unroll" and represent as a temporal process to be able to describe its complex structure. Thus, the "time" of the Bloch sphere could be different from our ordinary (classical) time, which is described by Einstein's relativity to be clear (although of course we cannot exclude this), but rather a useful *order parameter* that helps us to describe, as if it was a process, a structure that in its essence would not only be non-spatial, but also non-temporal.

Chapter 5

The Nature of Human Thought

Now that we have explained how to solve the probabilistic Bertrand's paradox and the quantum measurement problem, in this second half of the book we will take care of freeing the "third bird":[1] that of human cognition and of the problem of its modelization.

More precisely, the problem is about explaining why the traditional models, which are based on processes of the *logical-rational* kind and on the *classical theory of probabilities*, prove to be inadequate for modeling the results of numerous experiments of *cognitive psychology*. But not only that, it is also about explaining why the mathematical structure of quantum theory, and especially the quantum probabilities described by the Born rule, are so "unreasonably effective" in providing such modelization.

To better explicate what we are talking about, we begin by describing some of these situations that cannot be explained by the classical probability models. More precisely, we will consider the so-called *conjunction fallacy* and *disjunction effect*. A perfect example is the *Linda problem*, revealed in a brilliant experiment devised in 1983 by the Israeli psychologists *Amos Tversky* and *Daniel Kahneman* (see Figure 5.1).[2]

[1]See the book's subtitle.

[2]Kahneman was awarded the 2002 Nobel Prize in Economic Sciences for his studies of cognitive psychology applied to economics. Tversky would most probably have shared the prize, but he passed away in 1996.

Fig. 5.1 A. Tversky (left) and D. Kahneman (right).

The experiment is the following. A statistically relevant sample of people is presented with the following description of a woman called Linda (Morier & Borgida, 1984; Tversky & Kahneman, 1983):

31 years old, single, outspoken and very bright. She majored in philosophy. As a student, she was deeply concerned with issues of discrimination and social justice, and also participated in antinuclear demonstrations.

Based on this brief description, the subjects are then asked to evaluate the plausibility of the following statements:

(A) Linda is today active in a feminist movement;
(B) Linda is today a bank teller;
(C) Linda is today active in a feminist movement and is a bank teller;
(D) Linda is today active in a feminist movement or is a bank teller.

Once all the answers have been collected, one can calculate the average of the various evaluations, and obtain in this way the probabilities "P" associated with each of them. It is then observed that these obey the following relations (inequalities):

$$P(A) > P(A\,or\,B) > P(A\,and\,B) > P(B). \tag{5.1}$$

If we consider the first inequality, we have that according to the evaluation of the individuals subjected to the test, the probability $P(A)$ that Linda is today active in a feminist movement is strictly greater (>) than the probability $P(A\ or\ B)$ that Linda is today active in a feminist movement *or* is a bank teller. Similarly, if we consider the last inequality, we have that the probability $P(B)$ that Linda is today a bank teller is strictly lower (<) than the probability $P(A\ and\ B)$ that Linda is today active in a women's movement *and* that she is a bank teller.

This means that, according to the average opinion of the subjects, for Linda an alternative ($A\ or\ B$) is *less* probable than an absence of alternative (A), and that the concomitance of two events ($A\ and\ B$) is *more* probable than the occurrence of only one of the two events (B).

These are obviously contrary to the rules of classical probability. For this to be convincingly shown, it is sufficient to represent the events associated with the two possibilities by using *Venn diagrams* (see Figure 5.2).

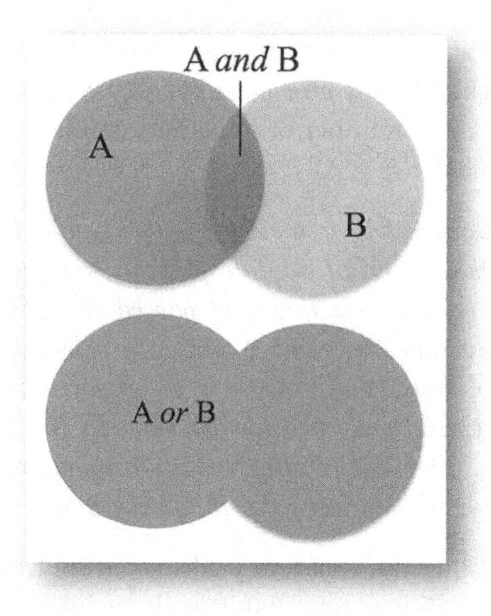

Fig. 5.2 The intersection "$A\ and\ B$" (above) and the union "$A\ or\ B$" (below).

Indeed, it follows from the axiomatic structure of classical probability theory that the probabilities are proportional to areas when the different possibilities are represented by sets. The *conjunction* of the two possibilities A and B is then equal to the *intersection* of the two representative sets, while the *disjunction* of the two possibilities A and B corresponds to the *union* of the two representative sets. Of course, since the area of the intersection of two sets is less than the areas of the individual sets, and the area of the union of two sets is greater than the areas of the individual sets, the correct inequalities, compatible with the theory of classical probability, are the following:

$$P(A) \leq P(A\, or\, B), \quad P(B) \leq P(A\, or\, B), \tag{5.2}$$

$$P(A) \geq P(A\, and\, B), \quad P(B) \geq P(A\, and\, B). \tag{5.3}$$

Thus, the Linda experiment violates the axiomatic rules of classical probability, and since there is an intimate relation between the latter and *Boolean logic*, the analysis shows that the outcomes of the Linda problem also violate classical logic.

Consider another example of an experimental situation that manifestly violates the *classical theory of rational choice*, designed by the same *Amos Tversky* and by the behavioral scientist *Eldar Shafir*, today known as the *Hawaii problem* (Tversky & Shafir, 1992). Tversky and Shafir asked a number of university students to consider the following two situations:

Situation 1: *Imagine that you have just taken a tough qualifying examination. It is the end of the fall quarter, you feel tired and run-down, and you are not sure that you passed the exam. In case you failed, you have to take the exam again in a couple of months, after the Christmas holidays. You now have an opportunity to buy a very attractive 5-day Christmas vacation package to Hawaii at an exceptionally low price. The special offer expires tomorrow, while the exam grade will not be available until the following day.*

You decide to: (i) buy the vacation package; (ii) not buy the vacation package; (iii) pay a $5 non-refundable fee in order to retain the rights to buy the vacation package at the same exceptional price the day after tomorrow, after you find out whether or not you passed the exam.

Situation 2: *Imagine that you have just taken a tough qualifying examination. It is the end of the fall quarter, you feel tired and run-down, and you find out that you passed the exam (alternatively, you find out that you failed the exam so that you will have to take it again in a couple of months, after the Christmas holidays). You now have an opportunity to buy a very attractive 5-day Christmas vacation package to Hawaii at an exceptionally low price. The special offer expires tomorrow.*

You decide to: (i) buy the vacation package; (ii) not buy the vacation package; (iii) pay a $5 non-refundable fee in order to retain the rights to buy the vacation package at the same exceptional price the day after tomorrow.

The choice of the students subjected to these two situations was as follows. More than 50% of them, in the situation in which they were aware of the outcome of the exam (54% in the case they knew they passed it, and 57% in case they knew they failed it) chose option (i), that is, they decided to buy the holiday package. However, only 32% of the students decided to buy it in the situation of uncertainty regarding the outcome of the examination (Situation 1).

In other words, if V denotes the event "I purchase the vacation package," E the event "I passed the exam," and $\bar{E}$ the event "I did not pass the exam," then, according to the experiment of Tversky and Shafir, the probability $P(V\, and\, E)$ of the joint event "$V\, and\, E$" is strictly greater than the probability $P(V)$ of the single event V; and the same holds for the probability $P(V\, and\, \bar{E})$, contrary to what is predicted by classical probability theory.

The effect identified in the Hawaii problem has been called the *disjunction effect*, since in the case of the two events E and $\bar{E}$, being mutually exclusive, their disjunction $E\, or\, \bar{E}$ is the event that is always certain (usually denoted Ω in probability theory). Indeed, it is certain that either I pass the exam or I don't pass it (*tertium*

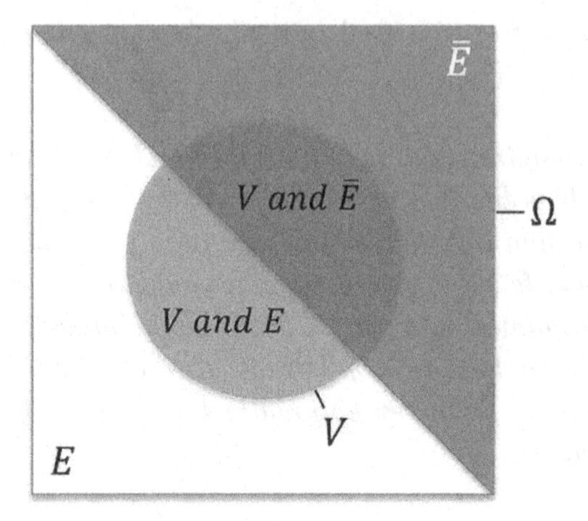

Fig. 5.3 The disjunction of the two mutually exclusive events E and $\bar{E}$, which corresponds to the union of the representative sets, gives the totality of the sample space Ω. The area of the intersection of the representative sets of V and E (or $\bar{E}$) is necessarily smaller than the area of V, and so are the corresponding probabilities.

non datur). So, event V and event "V *and* (E *or* $\bar{E}$)," that is, V considered concurrently with the *disjunction* "E *or* $\bar{E}$," are exactly the same event from the viewpoint of classical probability theory (see Figure 5.3). Yet, the uncertainty introduced by the presence of the disjunction appears to be able to affect the probability of V (hence, the name given to the effect), in the sense that we have the experimental inequalities:

$$P(V) = P(V\, and\, (E\, or\, \bar{E})) < P(V\, and\, E), \tag{5.4}$$

$$P(V) = P(V\, and\, (E\, or\, \bar{E})) < P(V\, and\, \bar{E}). \tag{5.5}$$

Thus, similar to the conjunction fallacy illustrated by the Linda problem, the disjunction effect also constitutes a flagrant violation of the axiomatic rules of classical probability and of classical Boolean logic, as can be readily and convincingly demonstrated if we consider the Venn diagram of Figure 5.3.

So, if we consider the results obtained in the problems of Linda and Hawaii, we realize that we humans, when we are in certain decision-making situations, we make, at least at first sight, wrong evaluations, which from a statistical point of view translate into specific probabilistic inequalities violating the classical probability theory. Why do we do this? The typical response is that we humans have little ability to manage situations of *uncertainty* and hence, when in such situations, become in part *irrational*. On the other hand, it would be useful to ask: what does it mean to think in an irrational way?

In mathematics, for example, we know that an *irrational number* is such because it cannot be written as a fraction of two integers. The discovery of irrational numbers is traditionally attributed to the Pythagorean *Hippasus of Metapontum*, who is said to have produced the first demonstration that the length of the hypotenuse of an isosceles right triangle, with catheti of length 1, could not be a *true* number, according to what was the (only rational) Pythagorean conception of numerical entities.

Indeed, for the Pythagoreans, the irrational numbers were true impossibilities and whose existence was unacceptable, so much so that, according to legend, they costed the life of poor Hippasus, who drowned along with his scandalous "secret."

In the case of human thought and decision-making processes, there is a certain similarity with the Pythagorean story. In fact, just as the existence of irrational numbers was in conflict with the beliefs of the Pythagoreans, the idea that the human cognition was also governed by processes not reducible to logic was likewise long ignored by scholars of the human mind, almost as if the irrational part of man was the expression of a defect, or a failure, of which it was better not to talk. Also, how could one talk about something that apparently did not have any structure? But, is this really so?

Consider again the analogy with irrational numbers. Whereas in the days of Pythagoras no one knew what they were exactly, or how many they were, today we know that they form the majority of the known numbers. In other words, whereas in ancient times it was believed that numbers such as $\sqrt{2}$ were only improbable anomalies, we now know that they are the rule, and that behind this rule there is a precise *mathematical structure* — that which precisely allows their construction as an extension of the set of rational numbers. And it is telling about today's conviction of the existence of irrational numbers that they have been simply denoted, along with rational ones, as *real numbers*.

The same thing, in a sense, is happening with human cognitive processes. For a long time it was thought that there was no clear structure in the irrational (typically subconscious, associative, intuitive) part of human thought, and that this reflected the exception that proved the rule of "true thought," typically conscious and logico-rational. Today, however, we know that the truth is exactly the opposite, i.e., that the irrational part is dominant and that it is not true that it is devoid of structure but that it is simply structured differently from the classic, analytical part.

Surprisingly, this structure is similar to the one that emerges from the interaction between microscopic quantum entities and macroscopic systems in charge of their measurement, in the sense that the probabilities obtained in experiments of cognitive psychology are structurally very similar to quantum probabilities described by the Born rule.

In other words, "irrationality" is not synonymous with "absence of meaning," but with "presence of a different modality of attribution of meaning." And this modality, remarkably, is of the quantum kind, as it is effectively described by the formalism of quantum mechanics, that is, by its mathematics, which was developed precisely in an attempt to describe experimental situations where the aspects of *contextuality* and *emergence* are dominant. The following question then arises:

Why are the very specific quantum probabilities predicted by the Born rule so "unreasonably effective" in modeling the experimental data obtained in the analysis of human cognition processes?

As we will explain, the notion of *universal measurement* will once again prove to be essential in solving this mystery, or at least part of it. But before we explain why, it will be useful to consider (in Chapter 6) a further example of an experimental situation, namely that of a study carried out by the cognitive psychologist *James A. Hampton* in the late eighties of the last century (Hampton, 1988). This situation fully reveals the similarity between human cognitive processes, which are about *conceptual entities*, and quantum processes, which are about *microscopic entities*.

According to the traditional view, which goes back to *Aristotle*, a concept is a kind of "container of objects": an ensemble of "things" that have one or more properties in common, and that at the same time differ in other specific characteristics. According to this idea, each concept is associated with a kind of predetermined rule, which allows determining whether or not a given exemplar — the name given to the object — is representative of the concept in question, that is, whether or not it is present in the container.

For example, if we consider the concept "white," we can assume that it describes the ensemble of all possible white objects that are present in our spatiotemporal theater. Similarly, the concept "cat" represents the ensemble of all objects that are cats, that is, carnivorous mammals belonging to the family of the Felidae, having specific anatomical features. When we combine the two concepts "cat" and "white," and form the composite concept "white cat," the idea is that this new concept describes the ensemble of exemplars obtained from the intersection of the two previous ensembles, that is, those exemplars that the "white" container and the "cat" container have in common (see Figure 5.4).

At first glance, this seems to work, and the temptation may be to assume that all the concepts are describable as abstract ensembles (containers) of exemplars, and that their combinations are always amenable to appropriate intersections of the corresponding ensembles. However, a moment's thought suffices to recognize that this representation is absolutely inadequate to capture the entire *potentiality of meanings* expressed by the various human concepts, that is, the way in which these are capable of generating new meanings in their possible combinations, when immersed in different cognitive contexts.

To demonstrate this, let us consider the two concepts "stone" and "lion." In this case, none of the exemplars representative of the first concept are also representative of the second concept (see Figure 5.5). In other words, the intersection of the two ensembles now corresponds to an *empty set*.

Fig. 5.4 The concept "white," understood as the ensemble of "objects of white color" (left), and the concept "cat," understood as the ensemble of "objects that are cats" (right). The intersection of these two ensembles can be understood as the representation of the concept "white cat," according to the compositional interpretation of the human concepts.

According to the above-mentioned *compositional interpretation*, the concept obtained from their combination should then be *semantically empty*, as no exemplar would correspond to it. On the other hand, we all know that the combination of these two concepts produces a new emergent concept, expressed by the term "stone lion," whose typical exemplars are simply stones shaped like lions, i.e., statues of lions (see the sub-ensemble of the left ensemble of Figure 5.5).

So, the hypothesis that the semantics of combinations is of the compositional kind is insufficient, as it doesn't have general validity, and one of the central problems of cognitive research is precisely that of identifying a model capable of describing the emergent meanings of all possible conceptual combinations.

Fig. 5.5 The intersection of the ensemble of exemplars of "stone" (left) and of the ensemble of exemplars of "lion" (right) is empty. Therefore, according to the compositional interpretation, the composite concept "stone lion" should be semantically empty.

In other words, it is about understanding how the meaning of a combination of concepts relates to the meaning of the concepts forming the combination. And in particular, it is about explaining how the different conceptual combinations are capable of producing effects of *overextension* and *underextension* of the probabilities, compared with the predictions of the classical theory, as already observed in the Hawaii and Linda problems.

To do this, we must provide a modeling of the conceptual entities that from the beginning highlights their contextual nature, i.e., their ability to change meaning depending on the semantic contexts. To return to the previous example, the concept "stone," in the context of "lion," becomes "stone lion," whose meaning is different from that of "stone."

In this case, the change is quite radical, but less radical changes are obviously possible. If we take the example of the concept "fruit," and we ask a sample of people to show us a representative exemplar of it, one of the most frequent answers will be "apple." On the other hand, if the term "fruit" is placed in the context of the concept "juicy" to get "juicy fruit," and we ask the same persons to give us a representative exemplar of "juicy fruit," the most frequent response will no longer be "apple" but "orange," "grape," or "watermelon," for example.

When a concept presents itself in the absence of combinations with other concepts, we can consider that it is in its *ground state*: a state in which it assumes its most *neutral* and *prototypical* meaning (or rather, set of possible meanings). Instead, when it is combined with other concepts, we can describe this process as a real *change of state*, that is, as the transition from the ground state to an *excited state*.[3]

In addition, if a concept is combined with one or more concepts in a premeditated way, we can consider that its change of state is the expression of a deterministic context, that is, a context capable of producing a predetermined outcome. In physics, this is the case of the famous *Schrödinger equation*, which describes the evolution of an isolated physical system, or of those processes that precede a measurement, called *preparations*.

When instead we ask a person to choose a typical exemplar of "fruit" or of "juicy fruit," the process is no longer fully under the conscious control of the mind that operates it, and different outcomes are *a priori* possible. Since, however, these outcomes are not predeterminable, the change produced by the decision-making process is the expression of an *indeterministic context*, as happens in a quantum measurement. Also in this case, of course, we have to do with a change of the state. In fact, "apple" is in turn a possible state of

[3]We are using here the typical quantum jargon, where an *excited state* of a physical system (such as an atom, a molecule, a nucleus, etc.) is any state that has a higher energy than the ground state, which is the state having the lowest possible energy.

Fig. 5.6 L. Gabora.

the concept "fruit," since "apple" can be equivalently described by the sentence (i.e., by the conceptual combination): "the fruit is an apple."

To describe human cognitive processes, we are using here language that is typical of modern physics, which makes use, precisely, of the fundamental notions of *state*, *property* and (experimental) *context* to describe systems. This was in fact the innovative idea that at the turn of the millennium one of us had — along with some colleagues, including the Canadian psychologist *Liane Gabora* (see Figure 5.6) — to describe human conceptual entities by using the language of quantum physics, the latter being a theory that was developed precisely in an attempt to describe (microscopic) entities whose interactive modalities are not only highly contextual, but often also totally unpredictable.

But it is time to enter into the merits of the previously mentioned experiment by Hampton and explain how this is equivalent to a measurement of the quantum kind, described by the Born rule. We will do this by comparing Hampton's experiment to the paradigmatic *double-slit experiment*, where between a light source and a detector screen (usually a photographic plate) an opaque barrier with two openings is positioned, able to generate a typical *interference pattern.* As we will see, it is possible to observe these *interference effects* not only in the elementary physical processes, but also in the human cognitive processes of the nonlogical-rational kind, responsible for the emergence of the non-classical probabilities described by the Born rule.

Chapter 6

Fruits Interfering with Vegetables

One of the things Hampton considered in his experiment was the composite concept "fruit or vegetable" (as we have seen, this can be considered to be either the conceptual entity "fruit" in a particular state or a new conceptual entity obtained by the combination of the two entities "fruit" and "vegetables" via the logical connector "or," which in turn, of course, is also a conceptual entity).

More precisely, Hampton submitted the following collection of possible exemplars of "food" to a total of 40 students in psychology at Stanford University (see Figure 6.1):

(1) almond, (2) acorn, (3) peanut, (4) olive, (5) coconut, (6) raisin, (7) elderberry, (8) apple, (9) mustard, (10) wheat, (11) ginger root, (12) chili pepper, (13) garlic, (14) mushroom, (15) watercress, (16) lentils, (17) green pepper, (18) yam, (19) tomato, (20) pumpkin, (21) broccoli, (22) rice, (23) parsley, (24) black pepper.

He then asked them to choose from the collection:

(1) a typical exemplar of "fruit;"
(2) a typical exemplar of "vegetable;"
(3) a typical exemplar of "fruit or vegetable."

This implies that the exemplars included in the collection will be chosen by the different subjects with different *relative frequencies* in relation to the three questions above, and we can associate to these

Fig. 6.1 The ensemble of 24 exemplars of "food" that are considered in Hampton's experiment.

relative frequencies the probabilities that human subjects will choose a given exemplar when subjected to one of these three questions.[1]

Let us consider some of the values obtained by Hampton. The probability (relative frequency) that "elderberry" is chosen as a typical exemplar of "fruit" is:

$$P(\text{fruit} \to \text{elderberry}) = 0.1138. \tag{6.1}$$

[1]In reality, Hampton did not perform exactly this experiment but a similar one consisting in asking each student to assess the *membership typicality* of each exemplar, in relation to the three concepts in question, according to a specific scale of values, called the *Likert scale*. This way of proceeding presents some statistical advantages and still allows making a good estimate of the probabilities in question, considering the strong similarity between a statistics derived from "estimating on a Likert scale each exemplar for its membership typicality" and from "choosing one typical exemplar." Of course, an experiment that directly collects the relative frequencies of "choosing one typical exemplar" could have been performed. Hampton did not do this experiment because he was investigating a connected issue, but different from the one we are interested in here. This means that we use Hampton's data in a way that was not meant to be, knowing however that if we had done the experiment ourselves for the data we use, these would be very close to the ones measured by Hampton.

Instead, the probability that "elderberry" is chosen as a typical exemplar of "vegetable" is:

$$P(\text{vegetable} \to \text{elderberry}) = 0.0170, \tag{6.2}$$

which is obviously much lower than the previous one. According to a classic interpretation, one would then expect that the probability $P(\text{fruit } \textit{or} \text{ vegetable} \to \text{elderberry})$, i.e., that "elderberry" is chosen as a typical exemplar of "fruit or vegetable," would be greater than both the probability $P(\text{fruit} \to \text{elderberry})$ and the probability $P(\text{vegetable} \to \text{elderberry})$. Instead, the obtained value is:

$$P(\text{fruit } \textit{or} \text{ vegetable} \to \text{elderberry}) = 0.0480. \tag{6.3}$$

We thus have a violation of the classical inequalities, since (see the Venn's diagrams in Figure 5.2, and the inequalities (5.2)–(5.3)):

$$P(\text{fruit} \to \text{elderberry}) > P(\text{fruit } \textit{or} \text{ vegetable} \to \text{elderberry}). \tag{6.4}$$

In other words, we observe an *underextension effect* of the probabilities, which leads to a violation of the classical inequalities.

One could argue that the human subjects, when subjected to choice (C), operate (albeit without being aware of this) according to a purely sequential logic: first they choose between (A) and (B); then, if they chose (A), they select a typical example of "fruit," and if they chose (B), they select a typical example of "vegetable." If the decision-making process is of this type, then the value of the probability $P(\text{fruit } \textit{or} \text{ vegetable} \to \text{elderberry})$ should correspond to the *arithmetic mean* of the two values $P(\text{fruit} \to \text{elderberry})$ and $P(\text{vegetable} \to \text{elderberry})$. On the other hand, since

$$\frac{0.1138 + 0.0170}{2} = 0.0654 > 0.0480, \tag{6.5}$$

in this case we also obtain an underextension effect of the experimental probability, compared to the predictions of the classic probabilistic calculus.

But Hampton's data also contain values corresponding to *overextension effects*. For example, in the case of the exemplar "mushroom," we have the following experimental probabilities:

$$P(\text{fruit} \to \text{mushroom}) = 0.0140, \quad (6.6)$$

$$P(\text{vegetable} \to \text{mushroom}) = 0.0545, \quad (6.7)$$

$$P(\text{fruit } \mathit{or} \text{ vegetable} \to \text{mushroom}) = 0.0604, \quad (6.8)$$

whereas the arithmetic mean of $P(\text{fruit} \to \text{mushroom})$ and $P(\text{vegetable} \to \text{mushroom})$ gives:

$$\frac{0.0140 + 0.0545}{2} = 0.0342 < 0.0604. \quad (6.9)$$

In other words, depending on the exemplars, when we consider the composite concept "fruit or vegetable," we obtain values for the probabilities that may be either lower or greater than those obtained by averaging the values of the probabilities of the individual (non-composite) concepts "fruit" and "vegetable."

These variations, both in the augmentative and in the diminutive sense compared to the average value, are typical in physics when we have to do with *interference effects.* And in fact, even in the case of Hampton's experiment, it is possible to represent the experimental data by bringing out a typical *interference pattern.*

To reveal this pattern, it is sufficient to use the mathematical quantum language in order to construct the quantum states ψ_A and ψ_B associated with "fruit" and "vegetable," respectively, from which it is possible to deduce, by the Born rule, the probabilities relative to the situations (A) and (B). The state $\psi_{A\,or\,B}$, associated instead with "fruit or vegetable," can be described as a *state of superposition* of the states ψ_A and ψ_B. If we think of ψ_A and ψ_B as *wave functions*, where the variables are the different exemplars, represented in a suitable "exemplars' space," it is clear that if $\psi_{A\,or\,B}$ is a superposition of ψ_A and ψ_B, this will be able to produce both constructive interferences, descriptive of overextension effects of the probabilities, and destructive interferences, descriptive of underextension effects of the probabilities (see Figure 6.2).

Before showing the specific interference pattern that one obtains in the analysis of Hampton's data, we open a brief parenthesis to recall what the American physicist *Richard Feynman* considered to

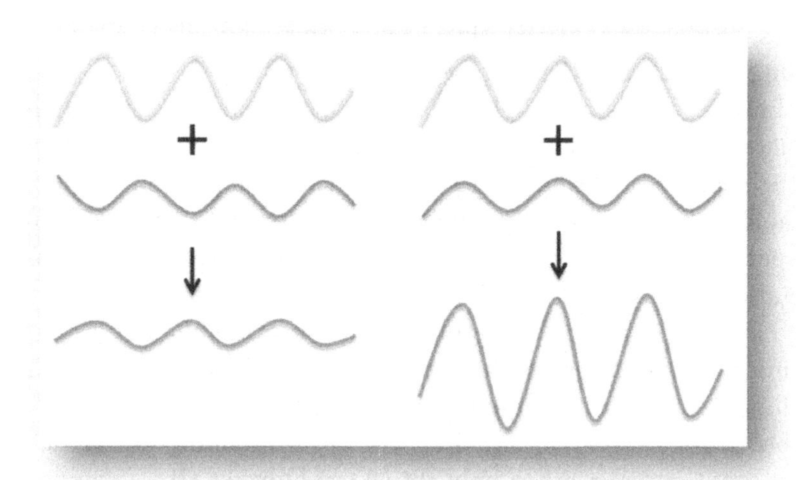

Fig. 6.2 In the superposition of two waves, both destructive and constructive interferences may occur, reducing the amplitude of the resultant wave (left), or increasing it (right), respectively.

be the key experiment for the understanding of quantum mechanics: that of the *double-slit* (which we already mentioned in the previous section). Indeed, the parallel between this and the experiment by Hampton will be particularly instructive.

In the double-slit experiment, a punctual light source, of very low intensity, emits *photons*,[2] one at a time, in the direction of an opaque barrier having two slits, which we shall denote (A) and (B). Behind the barrier, at a certain distance, is placed a screen (e.g., a photographic plate) able to detect the impacts produced by the individual photons coming from the two slits.

If the nature of the photons was corpuscular, we would observe on the detection screen a single area of impression, rather homogeneous, more intense where its surface receives the larger number of impacts, i.e., in correspondence of the two slits. As shown in Figure 6.3(a), one could also obtain this kind of pattern by proceeding in two separate

[2]The same experiment can also be carried out with elementary entities of the fermionic kind, like electrons for instance.

stages, first closing slit (B), thus leaving open only slit (A), and then closing slit (A), thus leaving open only slit (B).

In other words, if the photons were like corpuscles, i.e., tiny projectiles, the double-slit experiment would admit an interpretation of the compositional kind, in the sense that the pattern of impacts obtained when both slits are open would be entirely deductible from the patterns of impacts obtained when these are opened in succession, one after the other, rather than simultaneously.

This means, among other things, that the probability of observing an impact at a given point x of the screen, $P(A\,or\,B \to x)$, when both slits are open, is simply given by the arithmetic mean of the probabilities $P(A \to x)$ and $P(B \to x)$ of obtaining an impact at point x of the screen when only slit (A) or only slit (B), respectively, is open, i.e.:

$$P(A\,or\,B \to x) = \frac{P(A \to x) + P(B \to x)}{2}. \tag{6.10}$$

But the photons, even though they leave traces of point-like impacts on the photographic plate, exactly as if they were corpuscles, in fact are not corpuscles, and when they are subjected to the double-slit experiment, they produce a result that is not interpretable compositionally.

More precisely, when we use a source of photons in the situation where only one of the two slits is open, the distribution of impacts on the screen is perfectly compatible with the hypothesis that the photons are entities of a corpuscular nature. Instead, when the two slits are open at the same time, the distribution of impacts on the screen completely changes, being no longer deducible as a simple (weighted) sum of the impacts obtained when the two slits are opened separately from one another (see Figure 6.3(b) and Figure 6.4).

If we reason in probabilistic terms, there will be in this case some points x on the screen where the probability of observing a photon's impact will differ with respect to the value given by the arithmetic mean, in the sense that there will be points of *overexposure* on the screen, where the probability is greater, and points of *underexposure*, where the probability is lower.

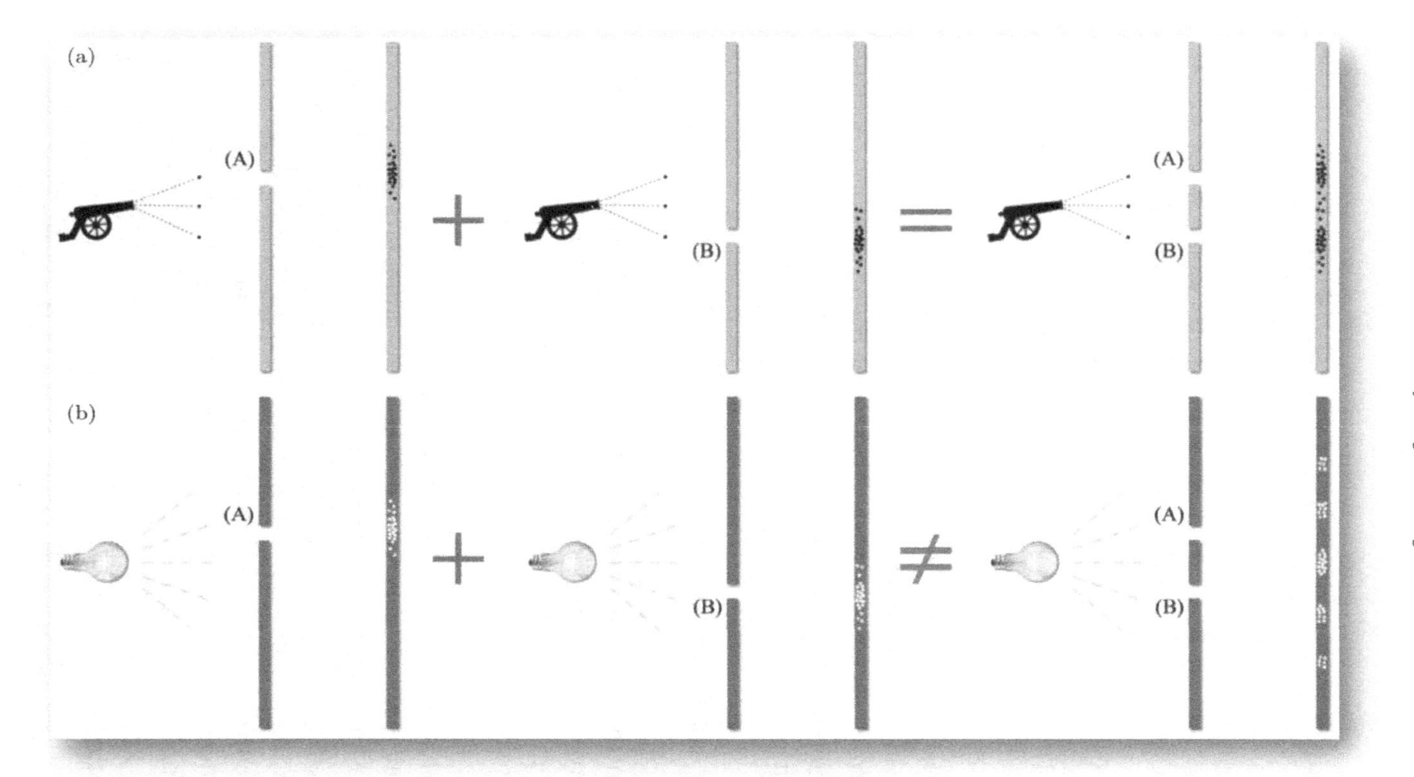

Fig. 6.3 The difference between (a) projectiles and (b) photons in single- versus double-slit situations. Projectiles and photons behave similarly when only one slit is open at a time, but in the case of photons the double-slit situation is not deducible from the single-slit ones.

Fig. 6.4 The typical fringe interference pattern observed on the final detector screen of a double-slit experiment.

In other words, it is necessary in this case to correct the arithmetic mean by introducing a third term, $I(x)$, called *interference contribution*, responsible for the effects of overextension (constructive interference) and underextension (destructive interference) of the probabilities:

$$P(A\, or\, B \to x) = \frac{P(A \to x) + P(B \to x)}{2} + I(x). \tag{6.11}$$

The attentive reader will have already grasped the profound analogy between the description of the double-slit experiment and the experiment by Hampton. To make this analogy even more stringent, we can reformulate the double-slit experiment in purely cognitive terms.

For this, we consider the detector screen as if it were a sort of *mind* that is able to answer questions. This mind, however, will not speak our human language, but a special language made up of "points on its surface." More precisely, points (photonic impacts) are the way by which the "screen-mind" communicates (and records) its answers.

Of course, to talk to this very particular mind, we must learn its language, that is, we have to know the meaning it attaches to the points that can appear in the different positions, and above all we must know the question the screen-mind is answering, with its pointillistic language. To this end, we assume that the photon emitted from the light source is an *abstract, conceptual entity*, and that the screen-mind is asked to choose a point that, according to its evaluation, better represents:

(1) a photon passing through slit "(A);"
(2) a photon passing through slit "(B);"
(3) a photon passing through slit "(A) or (B)."

Obviously, these three questions can be addressed to the screen-mind only in operational terms, i.e., by enacting them through the construction of a specific experimental context. Question (1) requires the presence of a barrier with a high positioned single slit; question (2), the presence of a barrier with a low positioned single slit; and question (3), the presence of a barrier having a double slit.

Now, put yourself in the place of the screen-mind, and try to answer the first two questions. Clearly, the answering process cannot be purely deterministic. In fact, when we say "passing through a slit," we are not specifying how exactly this happens. Moreover, when we say "passing through," we are implicitly saying that the photon in question will have to acquire, in the process, some *spatial properties*, otherwise the very notion of "passing through" would lose its meaning. But since there are many ways in which a spatial entity is able to pass through a slit, and nothing is specified about that in the question, as a screen-mind you will have the opportunity to choose from several possibilities.

This explains why every time the same question is asked, the answer (the point on the screen) can be different. On the other hand, the answer cannot be completely arbitrary, because the question specifies that the photon must pass, for example, through slit "(A)." So, with greater propensity the screen-mind will choose to respond by manifesting a point in a region which is located in proximity to said slit; and the same holds for question (2).

Things get more interesting when we consider question (3). In this case not only is there a level of uncertainty on how the photon will pass through either slit, but also about which of the two it will pass through. The screen-mind, confronted with this situation, will choose from among the possible answers those that best express this uncertainty. In other words, it will manifest a point-response that will be typical of a photon that passes through slit "(A)" or passes through slit "(B)," and when this same question is asked several times, the result will be the typical fringed structure shown in Figures 6.3(b) and 6.4.

Of course, it is not easy to get into the screen-mind and imagine being able to consciously reproduce such interference pattern. We are human minds, not screen-minds, and we do not speak the same language. Or rather, we have just recently learned how to decipher the quantum language of photons (and of other microscopic entities) and associated screen-minds (or other measuring instruments). On the other hand, as we will see, we humans are also able to create very complex interference patterns, actually even more complex than those produced by a screen-mind.

However, we can observe an interesting aspect in the structure of the answers represented in Figures 6.3(b) and 6.4. The fringe with a higher density of points is the central one, positioned at exactly equal distances from the two slits. So, the central fringe is where the screen-mind is most likely to manifest its answer when subjected to question (3). The reason is not hard to understand. When we observe the trace of an impact in the region of the central fringe, we find ourselves in a situation of maximum ignorance about the slit the photon would have used to pass through the barrier. In other words, the point-answers belonging to the central fringe best express this situation of doubt, and therefore constitute a perfect exemplification, in the form of points on the screen, of concept "(A) or (B)."

Now, the reasons why the screen-mind also selects the representative points of "(A) or (B)" in other regions, according to the typical "fringe structure," is of course more difficult to understand in cognitive terms, and one may think that the way a screen-mind is sensitive to the meaning of the above questions is ultimately very different from the way we humans attach meaning to human conceptual entities. But it is not necessarily so. In fact, as already stated, the screen-mind and the corresponding photonic language are much more structured than the human mind and the corresponding human language: for this reason, they are able to offer answers whose geometry is much more regular than that of human answers.

For a proper understanding of what do we mean, it is time to present all the data of Hampton's experiment. We will do this by pretending it were a kind of double-slit experiment. Of course, we are not dealing here, in the strict sense of the term, with a two-slit barrier, a light source and a detector placed behind the barrier, but the structure of the experimental context is very similar to this, and therefore we can represent it, symbolically, as if it was such.

Instead of the "photon," we are dealing with an abstract concept, for example the concept "food." Instead of slits "(A)" and "(B)," we have the concepts "fruit" and "vegetable," respectively; and instead of a photon passing through slit "(A)," slit "(B)," or double-slit "(A) or (B)," we have the situation of the concept "food" which is a "fruit," a "vegetable," or a "fruit or vegetable." And, of course, instead of the detector, we have the human minds of students, who instead of choosing a point on the screen choose one of the 24 exemplars of Figure 6.1.

As already mentioned, we can associate to each exemplar a coordinate in a special "space of exemplars," i.e., we can do as if the students, like the screen-mind, would respond to the three questions by indicating the coordinates of the exemplar that is chosen in each case on the exemplar screen. In this case, only 24 coordinates (i.e., points) can be chosen, and not every possible coordinate (i.e., point of the exemplar screen), but obviously this is not a fundamental problem, since the number of exemplars that can be considered is not *a priori* limited. Hampton considered 24 of them only for practical reasons.

Figures 6.5 and 6.6 illustrate the situation of question (1) and that of question (2), respectively, namely in the case where the exemplar chosen is typical of "fruit" and "vegetable", respectively.[3] The

[3]In Figure 6.5, and the following, the pattern of impacts produced by the source "food" on the exemplar detector screen was obtained by means of an appropriate interpolation of the values of the probabilities in the three different interrogative contexts, using, as is usual in quantum mechanics, a wave packet as a model of function. In this way, it becomes possible to assign a probability value (obtained by taking the square modulus of the wave packet) to each region of the screen, and not only to those corresponding to the 24 exemplars taken into account by the students.

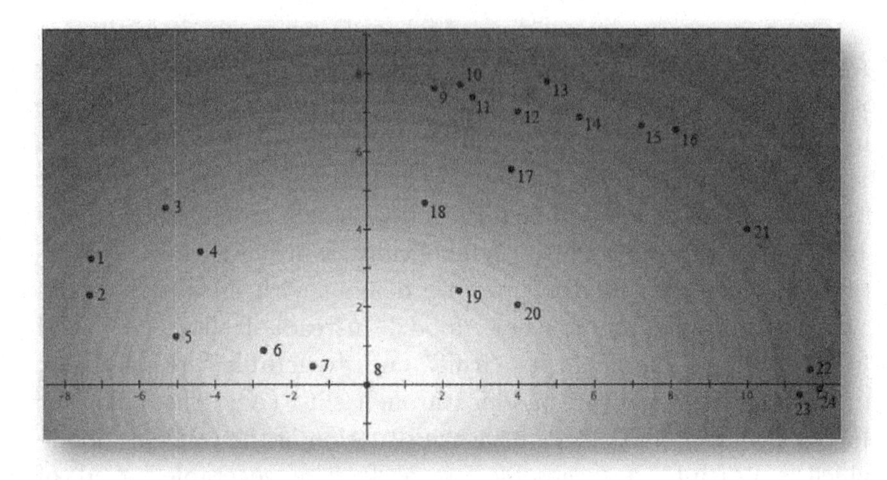

Fig. 6.5 The graphical representation of the experimental data obtained when the students are asked to choose (1) a typical exemplar of "fruit." The numbered points correspond to the specific coordinates of the 24 exemplars. The brighter is the region of the screen where a point is located, the larger is its probability of being selected (the number 8, which is at the origin of the chosen system of coordinates, is representative of the exemplar "apple").

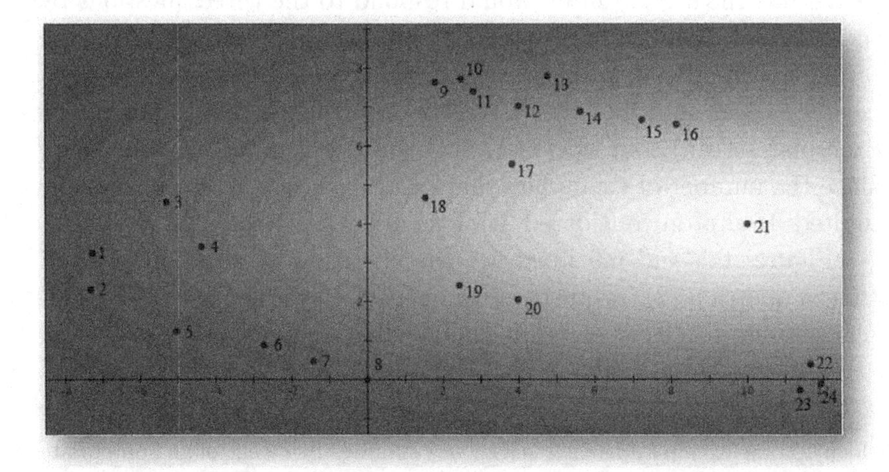

Fig. 6.6 The graphical representation of the experimental data obtained when the students are asked to choose (2) a typical exemplar of "vegetable." The brightest point on the screen, with number 21, is the point with the highest probability, and is representative of the exemplar "broccoli."

exemplar that is in the best correspondence with the "fruit" cognitive slit is "apple," which is at the origin of the coordinate system of the exemplar screen.

In fact, the probability that the exemplar "apple" is chosen as typical of "fruit" is very high: $P(\text{fruit} \rightarrow \text{apple}) = 0.1184$; for example, this is more than eight times as high as the probability $P(\text{fruit} \rightarrow \text{mushroom})$. The exemplar which instead is in the best correspondence with the "vegetable" cognitive slit is "broccoli," whose probability is: $P(\text{vegetable} \rightarrow \text{broccoli}) = 0.1284$, which for example is more than double the probability $P(\text{vegetable} \rightarrow \text{mushroom})$.

Now, if the concept "food," here interpreted as if it was a photon, behaved like a classical particle, we would expect the distribution of the impacts on the screen of the exemplars in the situation of question (3) to be simply the average of the distributions relative to the two questions (1) and (2), as illustrated in Figure 6.7 and Figure 6.9(a). But as we know, this is not the case, as the concept "fruit or vegetables" is not of the compositional kind.

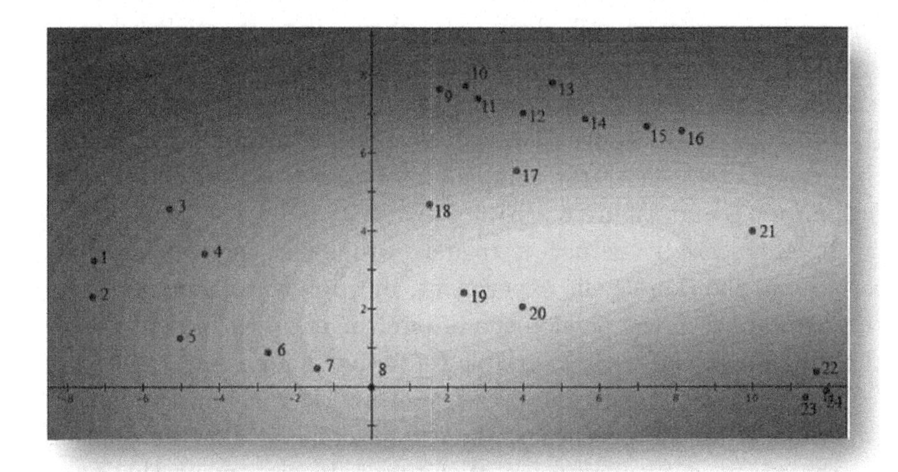

Fig. 6.7 If "food" would behave like a particle, with respect to the context of the "fruit or vegetable" cognitive double-slit, i.e., if "fruit or vegetable" was a conceptual combination of the compositional kind, then this would be the observed distribution of impacts on the screen of exemplars, which is typical of corpuscular entities.

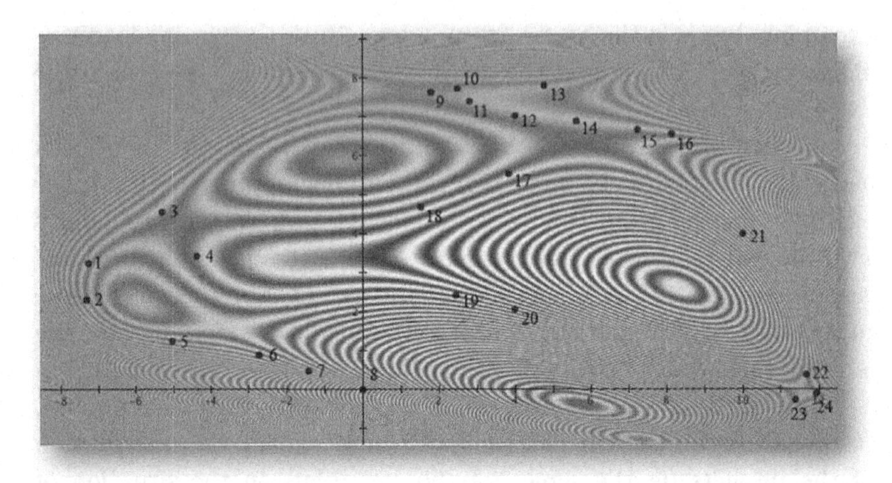

Fig. 6.8 The overextension and underextension effects of the probabilities observed by Hampton in his experiments result, in the quantum modeling of his data, in a particular interference pattern on the screen of the exemplars.

In the photonic metaphor, this corresponds to the fact that the photon-like entity "food" does not behave like a particle when it crosses the conceptual "fruit or vegetables" double slit. In fact, the collected experimental data (see Figure 6.10), modelized according to the principle of quantum superposition and the Born rule, give life, on the exemplar screen, to the complex interference pattern reproduced in Figure 6.8 and Figure 6.9(b).

In this case it is not a regular fringe-like interference pattern, as in the double-slit experiment, but a pattern reminiscent of those obtained in the phenomena of birefringence, when light passes through a material with a variable refractive index.

How can we explain the presence of interferences in Hampton's experiment? As noted earlier, these interferences allow accounting for the effects of overextension of the probabilities, when the interference is constructive, and of the effects of underextension of the probabilities, when the interference is destructive. But more specifically, what is the cognitive process that produces these effects, and why do some exemplars present overextension while others present underextension?

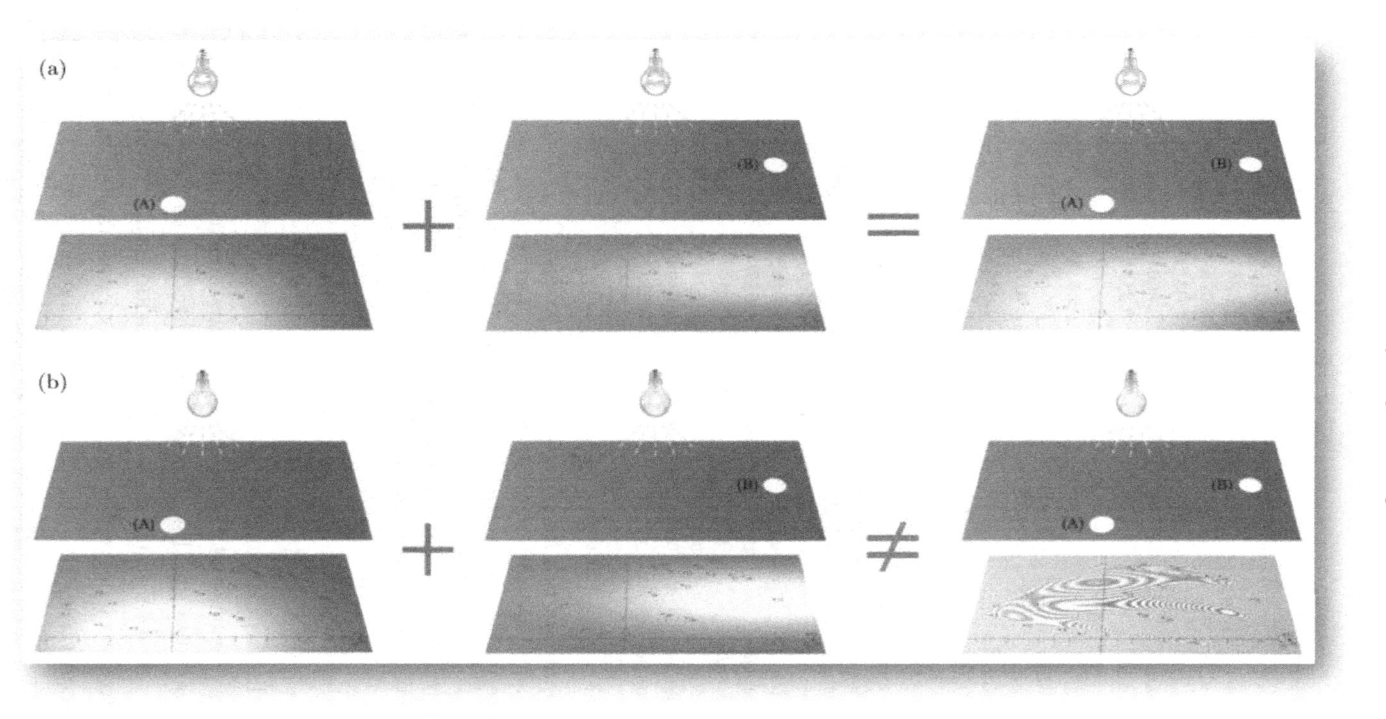

Fig. 6.9 (a) The hypothetical situation where the "fruit" and "vegetable" slits would be non-interfering alternatives, and (b) the real situation observed by Hampton with overextension and underextension (interference) effects. See also Figures 6.5–6.8.

Exemplars	P_f	P_v	$P_{f\,or\,v}$	$\bar{P}_{cl}$
(1) almond	0.0359	0.0133	**0.0269**	0.0246
(2) acorn	0.0425	0.0108	*0.0249*	0.0266
(3) peanut	0.0372	0.0220	*0.0269*	0.0296
(4) olive	0.0586	0.0269	*0.0415*	0.0428
(5) coconut	0.0755	0.0125	**0.0604**	0.0440
(6) raisin	0.1026	0.0170	*0.0555*	0.0598
(7) elderberry	0.1138	0.0170	*0.0480*	0.0654
(8) apple	0.1184	0.0155	**0.0688**	0.0670
(9) mustard	0.0149	0.0250	*0.0146*	0.0199
(10) wheat	0.0136	0.0255	*0.0165*	0.0195
(11) ginger root	0.0157	0.0323	**0.0385**	0.0240
(12) chili pepper	0.0167	0.0466	**0.0323**	0.0306
(13) garlic	0.0100	0.0301	**0.0293**	0.0200
(14) mushroom	0.0140	0.0545	**0.0604**	0.0342
(15) watercress	0.0112	0.0658	**0.0482**	0.0385
(16) lentils	0.0095	0.0713	*0.0338*	0.0404
(17) green pepper	0.0324	0.0788	*0.0506*	0.0556
(18) yam	0.0533	0.0724	*0.0541*	0.0628
(19) tomato	0.0881	0.0679	*0.0688*	0.0780
(20) pumpkin	0.0797	0.0713	*0.0579*	0.0755
(21) broccoli	0.0143	0.1284	*0.0642*	0.0713
(22) rice	0.0140	0.0412	*0.0248*	0.0276
(23) parsley	0.0155	0.0266	**0.0308**	0.0210
(24) black pepper	0.0127	0.0294	**0.0222**	0.0211

Fig. 6.10 Hampton's data for probabilities P_f, P_v, and $P_{f\,or\,v}$ associated with the different exemplars, when only the "fruit" slit, the "vegetable" slit, and both slits "fruit or vegetable," respectively, are open. The last column gives the classical average $\bar{P}_{cl} = \frac{P_f + P_v}{2}$, described in Figure 6.7. Note that in bold and italics are the probability values that are underextended and overextended, respectively, with respect to the corresponding classical values.

The explanation that we are going to put forward is similar to what we have indicated in the previous cognitive interpretation of the double-slit experiment.

Consider the case of the exemplar "mushroom." In general, the subjects, even though they consider mushrooms to be more representative of a vegetable than of a fruit, consider mushrooms not very representative of either category (the associated probabilities being very small, compared to "apple" or "broccoli," for example). On the other hand, we have seen that the probability that "mushroom" is chosen as a representative exemplar of "fruit or vegetable" is almost double compared to the arithmetic mean of the probabilities related to only "fruit" or "vegetable" (overextension effect).

To explain this deviation, we can consider that a human subject, when assessing the typicality of an exemplar in relation to the composite concept "fruit or vegetable," will proceed in a two-fold way — one that is both *logico-classical* and *quantum-conceptual*. The first way is about assessing the typicality of the exemplar in relation to its components "fruit" and "vegetable" taken separately, that is, by breaking down the concept into its components. This will produce a value compatible with the formula of the arithmetic mean.

The second way consists instead in considering "fruit or vegetable" as a new, *emerging* concept, not reducible, in regard to its meaning, to the meanings of its components taken individually. Therefore, in this second modality, the subject will evaluate if "mushroom" is an exemplar that can easily be separately attributed to "fruit" or "vegetables," and if this is not so, as it happens in this case, it will be assigned to the new emerging concept "fruit or vegetables." In other words, it will receive a very significant score according to this second modality of evaluation, resulting in an effect of overextension with respect to the classical evaluation (which only considers the first modality).

The same reasoning applies to the underextension effect, for example the one observed in the probability that "elderberry" is chosen as a typical exemplar of "fruit or vegetable." In this case, however, and contrary to "mushroom," it is not an exemplar that is difficult to classify separately as "fruit" or as "vegetable." Indeed, "elderberry" is considered to be a typical exemplar of "fruit." Therefore, it will receive a negative score as regards to its assignment to the emergent "fruit or vegetable" concept, thus producing a downward correction of the classical analytic-reductive evaluation.

We have explained how the traditional view of conceptual entities as mere "containers of exemplars" is unable to explain the many data collected, because the semantics of the conceptual combinations is generally non-compositional. We have also explained (within the limits of a non-mathematical presentation) how the quantum formalism, with its non-classical (non-Kolmogorovian) probabilities, is instead perfectly capable of doing so (for example, through the superposition principle). We can now state, and in part solve, the following riddle:

Why does the quantum rule of assignment of the probabilities, the so-called Born rule, prove to be so effective in modeling the experimental data of cognitive psychology?

Thanks to our solution of Bertrand's paradox and of the measurement problem, we now have everything we need to answer this question. To this end, we consider once again the model of the elastic membranes that we have discussed at length in the previous chapters. It is interesting to note how the breaking and collapse mechanism of the membrane constitutes a perfect metaphor for how we humans feel when we are confronted with a context of the decision-making kind. In other words, it is possible to understand the mechanism of the membrane also as the description of a psychological process of an intrapsychic nature.

Indeed, we can consider that when human beings are subjected to a question (or more generally, are urged to make a decision) which has N possible answers, they will automatically build a mental (neural) state of equilibrium that results from the balance of the different tensions between the initial state of the concept, subjected to the interrogative process, and the available answers, which, being mutually exclusive, compete with each other.

The elastic membrane can then be seen as an appropriate way to represent such mental state of equilibrium, characterized by the presence of N *tension lines* in mutual competition, which connect the point on the membrane representative of the initial state to the N vertices of the simplex representative of the possible answers (see Figure 6.11).

Always in agreement with what we can perceive in subjective terms, at some moment this mental equilibrium will be altered, in a non-predictable way, and this will trigger an irreversible process during which the initial conceptual state will be rapidly "attracted"

Fig. 6.11 The process of actualization of a potential answer, by a human subject confronted with an interrogative context, can be modeled as a *tension-reduction* process through the breaking of a membrane, described by a specific probability distribution ρ.

by one of the possible answers. This *tension-reduction* process is described in the extended Bloch model by the random breaking point of the membrane, which by collapsing will break the tensional equilibrium previously built.[4]

Now, every person subjected to a specific interrogative context will create a different mental state of equilibrium, determined by the uniqueness of her/his mindset, i.e., the specificity of the "conceptual network" that forms her/his memory structure. This means that each person will have a very personal *way of choosing an answer*, different from that of any other person.

[4]The *tension-reduction* process, however, does not always result in a full resolution of the conflict between the different possible answers. There are in fact interrogative contexts where the state of the system can be brought towards a further state of equilibrium, but relative to a smaller set of alternatives. This possibility is described in so-called *degenerate measurements*, which we have already mentioned but will not consider in this work.

Fig. 6.12 Each subject actualizes a response in a different way, each based on a different membrane, characterized by a different probability distribution ρ_i, $i = 1, 2, 3 \ldots$

In other terms, each individual will be associated with a specific membrane, characterized by a particular probability distribution (see Figure 6.12). And each membrane, as we know, is in turn associated with a given set of probabilities, for the different possible answers. This means that an experiment of cognitive psychology, involving a sufficient number of subjects, is by definition an experimental context in which an average of the universal kind is carried out, since the final statistics of the experiment involves a large number of *different ways of choosing an answer*, associated with the different participants.

In fact, the experimental probabilities associated with the different possible answers result from an average over the probabilities associated with all the individual subjects, that is, with all the membranes characteristics of their different *forma mentis*. Of course, the average produces a universal measurement only in the ideal limit of an infinite number of subjects, all different from each other, but what is important to understand here is that the universal average (associated with an effective uniform membrane) constitutes the best possible approximation when the experimenter is in a condition of maximum ignorance on how the different subjects choose. And since the Born rule corresponds precisely to a universal average, this in part explains why the quantum formalism is so well adapted to describe the experimental probabilities observed in the cognitive field.

If we say "in part" it is because the Born rule corresponds to a universal measurement that acts on a particular structure of the state space, which is that of the Hilbert space, specific to quantum mechanics. It is not certain, however, whether such Hilbertian structure will be strictly present in every cognitive experimental context.

In Hampton's experiment we have seen that the *superposition principle* was quite crucial to the modeling of the overextension and underextension effects, and such principle is characteristic of a *vector space* (the sum of two vectors being always a vector), as is the case of the Hilbert space of quantum mechanics. On the other hand, it is possible that in certain experimental situations such principle would apply only partially, and that therefore the rule of Born would not always be able to predict all the experimental probabilities.

In addition, it is possible that in some particular interrogative contexts, a violation of the quantum probabilities may be observed, and will undoubtedly be observed. This is either because the number of people involved is very small, or because not all the ways of choosing an answer are necessarily actualized by the participants for some given reasons.[5]

It is not the purpose of this short didactic text, however, to deal with these possible variations with respect to the Born rule, expression of the fact that the human thought processes are undoubtedly less structured than the microscopic quantum processes. They can be generally described as *quantum-like* processes, but not necessarily as *purely quantum* processes.

What we want to highlight here is that the notion of universal measurement, understood as a condition of mixture of all possible

[5]For example, some very special "meaning relations" among certain questions and the relative answers could advantage some "ways of choosing" and disadvantage some others, introducing a non-uniform average, thus a non-universal one.

measurements, that is, of mixture of all possible ways of answering a question, is perfectly natural in experimental situations that are typical of cognitive psychology, as these are the result of decision-making processes operated by many different subjects, each with a different mindset (thus, each with a different way of making choices), but who are not distinguished in the final statistics.[6]

And since quantum measurements, as we have seen, are universal measurements, we have now successfully released also the "third bird," that of the apparently unreasonable success of the Born rule outside of the domain of microphysics.

[6]It must be said that in principle even a single person may choose in different ways at different times, so producing a statistics of results that can be modelized in a first approximation by means of a universal average.

Chapter 7

Closing Thoughts

It is time to conclude our exploration of the notion of universal measurement, which allowed us to propose possible solutions to three fundamental problems: the long-standing Bertrand probability paradox; the central problem of quantum mechanics: the measurement problem; and the question of the surprising effectiveness of the Born rule in the experiments of cognitive psychology.

It is important to note that the idea of universal measurement (or average) has actually always been present, and indeed very visible, in that great theory of reality that is our human language. But in the words of *Edgar Allan Poe*, the best way to hide something is to put it in plain sight. Here, in a sense, we have taken what was in plain sight, and therefore so well hidden, and we have expressed it in a more precise way using the language of mathematics, and applied it (we think with success) to the three above-mentioned problems.

In several languages, in fact, an important distinction is made between two different forms of *lack of knowledge*. In Italian, for example, we have the term "aleatorio" (or the equivalent term "azzardo," i.e., "hazard" in English), indicating a danger, a risk, that is, something that can happen to us and that in no way we can control it. We find the same meaning for example in the Dutch term "toevallig," which comes from "fallen," which means "to fall," and "toe," which means "towards you." So: "that which falls towards us," which can "happen to us," such as losing money in a bet with nuts, that is, in a *game of chance* ("gioco d'azzardo," in Italian).

In Italian, there is however also the term "arbitrario" ("willekeurig" in Dutch, "arbitrary" in English), which designates "something that depends on the will or opinion of a single individual," such as when we decide to perform a particular experiment rather than another. In other words, our ancestors, at a time when language was developing, were already aware of the difference that existed between the randomness produced by the objects of their experience, expressed by the word "hazard," and the randomness produced by themselves, that is, by their actions, as subjects who are able to make choices, expressed by the word "arbitrary."

When we carry out an experiment, these two forms of randomness are always present, provided nothing intervenes to control them. The first form corresponds to the level of unpredictability associated with the experiment itself, once the protocol to be followed has been defined, while the second corresponds to the experimenter's lack of knowledge about the experiment s/he will actually choose (or has chosen) to perform.

Usually only the first type of randomness is taken into account in physics, both in conceptual and in formal terms. This is so because it is believed that the selection of the experiment to be performed is a process always under the full conscious control of the experimenter, and so, in that sense, it would not be random. But this is not always the case, as we have tried to highlight in the present essay.

The notion of universal measurement (and therefore of quantum measurement, when interpreted as a universal measurement) incorporates in itself these two distinct levels of lack of knowledge, or of randomness, already present in our natural language, but now expressed in a mathematically precise way in the formalism of the extended Bloch model.

What we have just observed raises one additional thought, which brings us back to the discussion presented in the preface to this book, when we evoked the notion of *robustness.* Although we can consider that an experimenter is totally unable to control the "dimension of randomness" inherent in a measurement (the breaking point of the membrane), certainly, if s/he wants to, s/he can take partly control of the process relative to the "dimension of arbitrariness" (the choice of the membranes). In the field of cognitive science, such control can be obtained via the choice of subjects having a particular *forma mentis*, that is, a particular "way of answering" when interrogated.

Of course, this is not what one will usually try to obtain, as this would affect the statistics of the outcomes in a way that would be more reflective of the experimenter's biased choice than of the reality of the observed entity. On the other hand, it is known that an experimenter, for example when making preparations to do an experiment in a physics laboratory, will proceed with a series of tunings and adjustments of the measuring apparatus. In doing so, s/he may operate a selection over the ρ-membranes that the experimental protocol will be able to actualize.

We can then ask whether this "adjustment process" is compatible with the desire to exert the slightest possible influence on the observed entity. Without going into technical details, it is possible to show (using the formalism of the extended Bloch model) that when an experimenter adjusts her/his instruments to achieve maximum *replicability* of the outcomes, that is, maximum *robustness* of their statistics with respect to small fluctuations of the entity's initial state, such adjustments correspond precisely to the situation of a universal measurement.

More precisely, the measurements that are maximally robust, that is, that offer the best guarantees in terms of replicability of the experimental results, are precisely those in which the experimenter exerts the least possible control over the choice of the membranes that the experiment is able to actualize. This also means that universal measurements correspond to experimental situations characterized by a form of randomness that we may call *immanent*, which is the remaining randomness after all attempts at control have been removed, that is, when one does not try, in any way, to reduce the level of indeterminism that is naturally present in the experimental context.

This definition of universal measurements, as experimental contexts in which the experimenter tries by all means to minimize their influence on how the different outcomes are produced, is fully compatible with their characterization in terms of a maximum lack of knowledge, which we have previously discussed. In fact, a maximum lack of knowledge also means a maximum lack of control; and a maximum lack of control means a minimum of influence on the observed entity.

Before we finish, we wish to refer the interested readers to several works that provide a more in-depth discussion of some of the concepts explored in this book, also from a technical point of view.

The first discussion of the notion of universal measurement can be found in two articles by one of us published in 1998 and 1999 (Aerts, 1998, 1999). Our proposed solution to the age-old Bertrand's paradox was published in 2015 in the *Journal of Mathematical Physics* (Aerts & Sassoli de Bianchi, 2015b).

Our possible solution to the measurement problem was published in 2014 in the *Annals of Physics* (Aerts & Sassoli de Bianchi, 2014), and it is based on earlier works of 1985 and 1986 (Aerts, 1985, 1986). In addition to this foundational text,[1] the authors jointly wrote other works on the extended Bloch representation and the corresponding hidden-measurements approach. We mention in particular an article about the more specific theme of the orientation of spins in space, or rather, on their lack of orientation (Aerts & Sassoli de Bianchi, 2015a).

In another work, written in the form of a dialog (Aerts & Sassoli de Bianchi, 2015c), the hidden-measurements approach (or interpretation) was compared with the "many-worlds" interpretation, proposed by Hugh Everett III in the fifties of the last century (DeWitt & Graham, 1973; Everett, 1957). Also, in (Aerts & Sassoli de Bianchi, 2016a), our approach was compared with that of the *possibilist*

[1]It would be more correct to say "re-foundational text," as dozens of works, by different authors, have been published in the past on the hidden-measurements approach. But what the approach was still missing to enter its "adulthood" was a natural model valid for each dimension N of the Hilbert space (i.e., for any number N of possible outcomes of a measurement process). Such model, which we have called the *extended Bloch model*, was presented for the first time only in (Aerts & Sassoli de Bianchi, 2014).

transactional interpretation of quantum mechanics (Kastner, 2013), showing that it provides a more complete description of the *weighted symmetry breaking* involved in a quantum measurement.

It is important to mention that the extended Bloch model also allows to shed light on another great quantum mystery, that of *entanglement.* In this regard, we point out the two articles (Aerts & Sassoli de Bianchi, 2015d, 2016b) where, among other things, we emphasize that the more complete version of quantum mechanics provided by the extended Bloch representation, in which the so-called *density operators* are also allowed to represent genuine states, does not only offer a possible solution to the measurement problem, but also to the lesser-known *entanglement problem.* This is because one no longer needs to give up the general physical principle saying that *a composite entity exists, and therefore is in a pure state, if and only if its components also exist, and therefore are in well-defined pure states.*

Finally, the proposed solution to the problem of the unreasonable effectiveness of quantum probabilities in the modeling of human cognitive processes was published in 2015 in two connected articles in the *Journal of Mathematical Psychology* (Aerts & Sassoli de Bianchi, 2015e,f) (see also (Aerts *et al.*, 2016; Aerts & Sassoli de Bianchi, 2016c)).

As stated in the Preface, one of us (Diederik) is a pioneer in the new and booming research field of *quantum cognition*, and in the above articles the reader will find numerous references to the most important publications in this ambit; see in particular (Busemeyer & Bruza, 2012; Haven & Khrennikov, 2013; Khrennikov, 2010; Wendt, 2015). Inspired by the successes of quantum cognition, Aerts also proposed in 2009 a new interpretation of quantum mechanics, fully compatible with the hidden-measurements interpretation, called the *conceptuality interpretation.*

Based on the observation that the formalism of quantum mechanics is able to model the human concepts so well, the hypothesis underlying this interpretation is that quantum particles (which particles are not) would in turn be *entities of a conceptual type.* More precisely, the hypothesis at the basis of the conceptuality interpretation is that the *nature* of a quantum entity is conceptual in the sense that it interacts with a measuring apparatus (or with an entity made of ordinary matter) in an analogous way as a concept interacts with the human mind (or with an arbitrary memory structure sensitive to concepts).

In other terms, similar to the "conceptual way" we used to explain the double-slit experiment, the elementary microscopic entities, whilst not describable as particles, waves or fields (since they are not representable in the three-dimensional space, or four-dimensional spacetime), would nevertheless behave as something that is very familiar to us all, and that we continually experience in a very intimate and direct way: *concepts*.

Of course, it is not within the scope of this book to present all the subtleties and complexities of the explanatory framework offered by this interpretation, and its effectiveness in explaining quantum phenomena such as entanglement and non-locality, which are traditionally considered to be "not understood" or "not understandable." We therefore leave to the reader the intellectual pleasure of discovering these explanations directly from the foundational work of their author: (Aerts, 2009, 2010a,b,c, 2014).

But to avoid any misunderstanding, let us emphasize that although the conceptuality interpretation suggests that quantum entities are *concepts*, and not *objects*, the conceptual entities associated with microscopic "particles" must not be confused with *human concepts*. Quantum entities would be conceptual in the sense that the notion that gives rise to the "way of being" (the "beingness") of a quantum entity and of a human concept are the same, as for example the notion of "wave" can describe both the mode of being of an electromagnetic wave and that of a sound wave. But other than that, they remain very different entities, much like an electromagnetic wave and a sound wave are very different entities.

We conclude this brief "reading guide" by indicating a very informative booklet (only available in English as an e-book), written by one of us (Massimiliano Sassoli de Bianchi), where some of the concepts that we have discussed in the present work are fully explained and illustrated: (Sassoli de Bianchi, 2013).

Lastly, for the "sons of the Internet," we point out two video presentations available on YouTube: (Sassoli de Bianchi, 2012; Sassoli de Bianchi & Sassoli de Bianchi, 2015). In the second video, there are some captivating animations of the unfolding of a measurement process within the Bloch sphere, i.e., animations that allow one to visualize what usually, and erroneously, is believed to be impossible to visualize!

Bibliography

Aerts, D. (1985). A possible explanation for the probabilities of quantum mechanics and a macroscopical situation that violates Bell inequalities. In: Mittelstaedt, P. and Stachow, E. W. (Eds.). *Recent Developments in Quantum Logic; Grundlagen der Exacten Naturwissenschaften*, Vol. 6, 235–251. Mannheim, Bibliographisches Institut.

Aerts, D. (1986). A possible explanation for the probabilities of quantum mechanics. *J. Math. Phys.* **27**, 202–210.

Aerts, D. (1998). The entity and modern physics: The creation discovery view of reality. In E. Castellani (Ed.), *Interpreting Bodies: Classical and Quantum Objects in Modern Physics*, pp. 223–257. Princeton: Princeton Unversity Press.

Aerts, D. (1999). The stuff the world is made of: Physics and reality. In D. Aerts, J. Broekaert and E. Mathijs (Eds.), *Einstein Meets Magritte: An Interdisciplinary Reflection*, pp. 129–183. Dordrecht: Springer Netherlands.

Aerts, D. (2009). Quantum particles as conceptual entities: A possible explanatory framework for quantum theory. *Foundations of Science* **14**, 361–411.

Aerts, D. (2010a). Interpreting quantum particles as conceptual entities. *International Journal of Theoretical Physics* **49**, 2950–2970.

Aerts, D. (2010b). A potentiality and conceptuality interpretation of quantum physics. *Philosophica* **83**, 15–52.

Aerts, D. (2010c). Quantum theory and conceptuality: Matter, stories, semantics and space-time. *Scientiae Studia* **11**, 75–100.

Aerts, D. (2014). Quantum theory and human perception of the macro-world. *Frontiers in Psychology* **5**, 554, doi: 10.3389/fpsyg.2014.00554.

Aerts, D. & Sassoli de Bianchi, M. (2014). The extended bloch representation of quantum mechanics and the hidden-measurement solution to the measurement problem. *Annals of Physics* **351**, 975–1025. See also the Erratum: *Annals of Physics* **366**, 197–198 (2016).

Aerts, D. & Sassoli de Bianchi, M. (2015a). Do spins have directions? *Soft Comput.*, doi:10.1007/s00500-015-1913-0.

Aerts, D. & Sassoli de Bianchi, M. (2015b). Solving the hard problem of bertrand's paradox. *J. Math. Phys.* **55**, 083503.

Aerts, D. & Sassoli de Bianchi, M. (2015c). Many-measurements or many-worlds? A dialogue. *Found. Sci.* **20**, 399–427.

Aerts, D. & Sassoli de Bianchi, M. (2015d). The extended bloch representation of quantum mechanics. explaining superposition, interference and entanglement. To be published in the *J. Math. Phys.* ArXiv:1504.04781 [quant-ph].

Aerts, D. & Sassoli de Bianchi, M. (2015e). The unreasonable success of quantum probability I: Quantum measurements as uniform measurements. *J. Math. Psych.* **67**, 51–75.

Aerts, D. & Sassoli de Bianchi, M. (2015f). The unreasonable success of quantum probability II: Quantum measurements as universal measurements. *J. Math. Psych.* **67**, 76–90.

Aerts, D. & Sassoli de Bianchi, M. (2016a). Quantum measurements as weighted symmetry breaking processes: The hidden measurement perspective. To be published in the *International Journal of Quantum Foundations.* ArXiv:1601.05222 [quant-ph].

Aerts, D. & Sassoli de Bianchi, M. (2016b). A possible solution to the second entanglement paradox. In: Aerts, D., De Ronde, C., Freytes, H. and Giuntini, R. (Eds.). *Probing the Meaning of Quantum Mechanics. Superpositions, Dynamics, Semantics and Identity*, pp. 351–359. World Scientific Publishing Company, Singapore.

Aerts, D. & Sassoli de Bianchi, M. (2016c). The GTR-model: A universal framework for quantum-like measurements. In: Aerts, D., De Ronde, C., Freytes, H. and Giuntini, R. (Eds.). *Probing the Meaning of Quantum Mechanics. Superpositions, Dynamics, Semantics and Identity*, pp. 91–140. World Scientific Publishing Company, Singapore.

Aerts, D., Sassoli de Bianchi, M. & Sozzo, S. (2016). On the foundations of the Brussels operational-realistic approach to cognition. *Frontiers in Physics* **4**. doi: 10.3389/fphy.2016.00017.

Bernard, C. (1949). *An Introduction to the Study of Experimental Medicine.* Paris: Henry Schuman, Inc.

Bertrand, J. (1889). *Calcul des Probabilités.* Gauthier-Villars et fils, Paris.

Busemeyer, J. R. & Bruza, P. D. (2012). *Quantum Models of Cognition and Decision.* New York: Cambridge University Press.

DeWitt, B. & Graham N. (eds.) (1973). *The Many-Worlds Interpretation of Quantum Mechanics.* Princeton University Press, Princeton.

Everett, H. (1957). Relative state formulation of quantum mechanics. *Review of Modern Physics* **29**, 454–462.

Hampton, J. A. (1988). Disjunction of natural concepts. *Memory & Cognition* **16**, 579–59.

Haven, E. & Khrennikov, A.Y. (2013). *Quantum Social Science.* Cambridge, Cambridge University Press.

Jaynes, E. T. (1973). The well-posed problem. *Foundations of Physics* **3**, 477–493.

Karimi, E. *et al.* (2012). Spin-to-orbital angular momentum conversion and spin-polarization filtering in electron beams. *Phys. Rev. Lett.* **108**, 044801.

Kastner R. E. (2013). *The Transactional Interpretation of Quantum Mechanics: The Reality of Possibility.* Cambridge University Press, New York.

Keynes, J. M. ([1921] 1963). *A Treatise on Probability.* London: Macmillan.

Khrennikov, A. (2010). *Ubiquitos Quantum Structure: From Psychology to Finance.* New York: Springer.

Kolmogorov, A. N. (1933). *Grundbegriffe der Wahrscheinlichkeits-rechnung.* J. Springer, Berlin.

London, F. & Bauer, E. (1939). *Expos de Physique Générale III*, pp. 1–51. Hermann, Paris.

Morier, D. M. & Borgida, E. (1984). The conjunction fallacy: A task specific phenomena? *Personality and Social Psychology Bulletin* **10**, 243–252.

Rowbottom, D. P. (2013). Bertrand's paradox revisited: Why Bertrand's 'solutions' are all inapplicable. *Philosophia Mathematica (III)* **21**, 110–114.

Sassoli de Bianchi, M. (2012). *Heisenbergs Uncertainty Principle and Quantum Non-Spatiality* (*Non-locality*) [Video file]. Retrieved from: `https://youtu.be/9C3vtVADL1o`.

Sassoli de Bianchi, M. (2013). *Observer Effect — The Quantum Mystery Demystified.* Adea edizioni, Sesto ed Uniti.

Sassoli de Bianchi, M. (2016). Theoretical and conceptual analysis of the celebrated 4π-symmetry neutron interferometry experiments. *Found Sci.*, doi:10.1007/s10699-016-9491-x.

Sassoli de Bianchi, M. & Sassoli de Bianchi, L. (2015). *Solving the Measurement Problem* [Video file]. Retrieved from: `https://youtu.be/Tk4MsAfC8vE`.

Shackel, N. (2007). Bertrand's paradox and the principle of indifference. *Philos. Sci.* **74**, 150–175.

Stapp, H. P. (2011). *Mindful Universe. Quantum Mechanics and the Participating Observer.* The Frontiers Collection. Springer-Verlag, Berlin Heidelberg, 2nd edition.

Tversky, A. & Kahneman, D. (1983). Extension versus intuitive reasoning: The conjunction fallacy in probability judgment. *Psychological Review* **90** (4), 293–315.

Tversky, A. & Shafir, E. (1992). The disjunction effect in choice under uncertainty. *Psychological Science* **3** (5), 305–309.

Von Neumann, J. (1932). *Mathematische Grundlagen der Quanten-mechanik.* J. Springer, Berlin.

Wendt, A. (2015). *Quantum Mind and Social Science.* Cambridge, Cambridge University Press.

Wigner, E. (1961). Remarks on the mind-body problem. In: Good, I. J. (Ed.). *The Scientist Speculates*, pp. 284–302. Heinemann, London.

Index

Printed in the United States
By Bookmasters